Erwin Dee Kord (Ed.)

Delaware Route 44

Erwin Dee Kord (Ed.)

Delaware Route 44

Delaware Route 300, Delaware Route 8, Delaware Route 11, Delaware Route 8, Kent County, Delaware

Solv

Contents

Delaware_Route_44

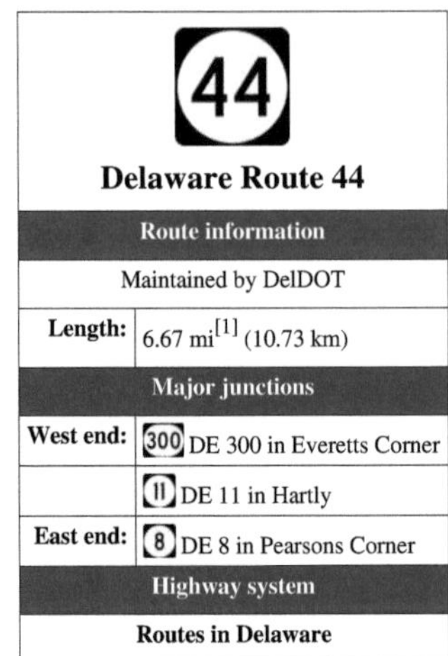

Delaware Route 44 is a state highway in Kent County, Delaware. It runs from Delaware Route 300 in Everetts Corner to Delaware Route 8 in Pearsons Corner, passing through the town of Hartly. The route was built as a state highway east of Hartly by 1924 and west of Hartly by 1932, receiving the DE 44 designation by 1936.

Route description

Delaware Route 44 heads to the southeast of DE 300 on Everetts Corner Road. It passes through a mix of woodland and farmland before reaching the town of Hartly, where it becomes Main Street. In Hartly, the route passes by homes and intersects Delaware Route 11. It then heads to the east out of Hartly on Hartly Road, passing through more rural areas. The route continues to its eastern termiuns at Delaware Route 8 near Pearsons Corner.[1] [2]

History

By 1920, what is now DE 44 existed as an unimproved county road.[3] The road east of Hartly was improved into a state highway by 1924.[4] By 1932, the portion of road west of Hartly became a state highway.[5] When Delaware created its state highway system by 1936, DE 44 was assigned to its current alignment.[6] The route has not changed since its inception.[7]

Major intersections

The entire route is in Kent County.

Location	Mile[1]	Road	Notes
Everetts Corner	0.00	300 DE 300 (Sudlersville Road)	Western terminus
Hartly	2.89	11 DE 11 (Arthursville Road)	
Pearsons Corner	6.67	8 DE 8 (Halltown Road/Forrest Avenue)	Eastern terminus
1.000 mi = 1.609 km; 1.000 km = 0.621 mi			

References

[1] http://www.deldot.gov/information/pubs_forms/manuals/traffic_counts/2006/pdf/rpt_pgs1_38_rev.pdf DelDOT 2006 Traffic Count and Mileage Report

[2] Google, Inc. *Google Maps — overview of Delaware Route 44* (http://maps.google.com/maps?f=d&source=s_d&saddr=delaware+44+and+delaware+300&daddr=delaware+8+and+delaware+44&hl=en&geocode=&mra=ls&sll=39.182605,-75.71168&sspn=0.114433,0.338173&ie=UTF8&t=h&z=13) (Map). Cartography by Google, Inc. . Retrieved 2010-08-20.

[3] Delaware Department of Transportation (PDF). *Delaware Official Highway Map* (http://www.deldot.gov/archaeology/historic_pres/historic_highway_maps/pdf/cd_002.pdf) (Map) (1920 ed.). . Retrieved 2010-04-15.

[4] Delaware Department of Transportation (PDF). *Delaware Official Highway Map* (http://www.deldot.gov/archaeology/historic_pres/historic_highway_maps/pdf/cd_003.pdf) (Map) (1924 ed.). . Retrieved 2010-08-01.

[5] Delaware Department of Transportation (PDF). *Delaware Official Highway Map* (http://www.deldot.gov/archaeology/historic_pres/historic_highway_maps/pdf/cd_006.pdf) (Map) (1932 ed.). . Retrieved 2010-08-20.

[6] Delaware Department of Transportation (PDF). *Delaware Official Highway Map* (http://www.deldot.gov/archaeology/historic_pres/historic_highway_maps/pdf/cd_008.pdf) (Map) (1936/37 ed.). . Retrieved 2010-04-15.

[7] Delaware Department of Transportation (PDF). *Delaware Official Highway Map* (http://www.deldot.gov/archaeology/historic_pres/historic_highway_maps/pdf/cd_083.pdf) (Map) (2008 ed.). . Retrieved 2010-08-02.

Kent_County,_Delaware

<table>
<tr><td colspan="2" align="center">Kent County, Delaware</td></tr>
<tr><td colspan="2" align="center">
The Kent County Courthouse in Dover in 2006.</td></tr>
<tr><td colspan="2">
Seal</td></tr>
<tr><td colspan="2" align="center">
Location in the state of Delaware</td></tr>
<tr><td colspan="2" align="center">
Delaware's location in the U.S.</td></tr>
<tr><td>Founded</td><td>1683</td></tr>
<tr><td>Seat</td><td>Dover</td></tr>
<tr><td>Largest city</td><td>Dover</td></tr>
</table>

Area	800.12 sq mi (2072 km²)
- Total	589.72 sq mi (1527 km²)
- Land	210.40 sq mi (545 km²), 26.30%
- Water	
Population	162,310
- (2010)	275/sq mi (106.2/km²)
- Density	
Congressional district	At-large
Website	www.co.kent.de.us [1]

Kent County is a county located in the central part of the U.S. state of Delaware. It is coextensive with the Dover, Delaware, Metropolitan Statistical Area. As of 2010 the population was 162,310, a 28.1% increase over the previous decade.[2] The county seat is Dover, the state capital. It is named for Kent, an English county.

History

In about 1670 the English began to settle in the valley of the St. Jones River earlier known as Wolf Creek. On June 21, 1680, The Duke of York chartered St. Jones County, which was carved out of New Amstel/New Castle County and Hoarkill/Sussex County. St. Jones County was transferred to William Penn on August 24, 1682, and became part of Penn's newly chartered Delaware Colony.[3]

Penn ordered a court town to be laid out, and the courthouse was built in 1697. The town of Dover, named after the city of Dover in England's Kent, was finally laid out in 1717, and became the capitol of Delaware in 1777. In 1787 Delaware was first to ratify the U.S. Constitution, and became "the First State." Kent County was a small grain farming region in the 18th Century.

More recently, in the 1960s, Dover was the scene of the manufacturing of the spacesuits worn by NASA astronauts in the Apollo moon flights by ILC Dover, now based in the small town of Frederica. The suits, dubbed the "A7L," was first flown on the Apollo 7 mission in October 1967, and was the suit worn by Neil Armstrong and Buzz Aldrin on the Apollo 11 mission. The company still manufactures spacesuits to this day—the present-day Space Shuttle "soft" suit components (the arms and legs of the suit).

Geography

According to the 2000 census, the county has a total area of 800.12 square miles (2072.3 km^2), of which 589.72 square miles (1527.4 km^2) (or 73.70%) is land and 210.40 square miles (544.9 km^2) (or 26.30%) is water.[4]

Adjacent counties

- New Castle County, Delaware - north
- Salem County, New Jersey - northeast[1]
- Cumberland County, New Jersey - east[1]
- Cape May County, New Jersey - east[1]
- Sussex County, Delaware - south
- Caroline County, Maryland - southwest
- Queen Anne's County, Maryland - west
- Kent County, Maryland - west

[1] *across Delaware Bay; no land border*

The border with Kent County, Maryland is rather unusual since the counties both share the same name. The only other instances in which two neighboring counties with the same name share a state border are Sabine County, Texas and Sabine Parish, Louisiana, Bristol County, Massachusetts and Bristol County, Rhode Island, Union County, Arkansas and Union Parish, Louisiana, and Escambia County, Alabama and Escambia County, Florida respectively.

National protected area

- Bombay Hook National Wildlife Refuge

Demographics

Historical populations		
Census	Pop.	%±
1790	18920	—
1800	19554	3.4%
1810	20495	4.8%
1820	20793	1.5%
1830	19913	−4.2%
1840	19872	−0.2%
1850	22816	14.8%
1860	27804	21.9%
1870	29804	7.2%
1880	32874	10.3%
1890	32664	−0.6%
1900	32762	0.3%
1910	32721	−0.1%
1920	31023	−5.2%
1930	31841	2.6%
1940	34441	8.2%
1950	37870	10.0%
1960	65651	73.4%
1970	81892	24.7%
1980	98219	19.9%
1990	110993	13.0%
2000	126697	14.1%
2010	162310	28.1%
[5] [6] [7]		

As of the census[8] of 2000, there were 126,697 people, 47,224 households, and 33,623 families residing in the county. The population density was 215 people per square mile (83/km²). There were 50,481 housing units at an average density of 86 per square mile (33/km²). The racial makeup of the county was 73.49% White, 20.66% Black or African American, 0.64% Native American, 1.69% Asian, 0.04% Pacific Islander, 1.27% from other races, and 2.22% from two or more races. 3.21% of the population were Hispanic or Latino of any race. 13.3% were of German, 11.3% United States or American, 10.9% Irish, 10.0% English and 5.4% Italian ancestry according to

Census 2000. 92.5% spoke English and 3.3% Spanish as their first language.

There were 47,224 households out of which 35.50% had children under the age of 18 living with them, 52.90% were married couples living together, 13.80% had a female householder with no husband present, and 28.80% were non-families. 23.00% of all households were made up of individuals and 8.40% had someone living alone who was 65 years of age or older. The average household size was 2.61 and the average family size was 3.06.

In the county the population was spread out with 27.30% under the age of 18, 10.10% from 18 to 24, 29.80% from 25 to 44, 21.20% from 45 to 64, and 11.70% who were 65 years of age or older. The median age was 34 years. For every 100 females there were 93.10 males. For every 100 females age 18 and over, there were 89.60 males.

The median income for a household in the county was $40,950, and the median income for a family was $46,504. Males had a median income of $32,660 versus $24,706 for females. The per capita income for the county was $18,662. About 8.10% of families and 10.70% of the population were below the poverty line, including 14.80% of those under age 18 and 8.80% of those age 65 or over.

Communities

Cities

- Dover
- Harrington
- Milford (part of Milford is in Sussex County)

Towns

- Bowers
- Camden
- Cheswold
- Clayton (part of Clayton is in New Castle County)
- Farmington
- Felton
- Frederica
- Hartly
- Houston
- Kenton
- Leipsic
- Little Creek
- Magnolia
- Smyrna (part of Smyrna is in New Castle County)
- Viola
- Woodside
- Wyoming

Census-designated places

- Dover Air Force Base (a CDP as "Dover Base Housing")
- Highland Acres
- Kent Acres
- Rising Sun-Lebanon
- Riverview
- Rodney Village
- Woodside East

Other localities

- Andrewville
- Berrytown
- Little Heaven
- Marydel

Trivia

Kent County, Delaware, is one of the few counties in the United States to border its namesake in another state—in this case, Kent County, Maryland.

See also

- National Register of Historic Places listings in Kent County, Delaware

References

[1] http://www.co.kent.de.us

[2] http://www.delawareonline.com/article/20110302/NEWS/110302038/
New-census-figures-released-Delaware?odyssey=tab|topnews|text|Home

[3] *NEW YORK: Atlas of Historical County Boundaries* by John H. Long and Kathryn Ford Thorne

[4] "Census 2000 U.S. Gazetteer Files: Counties" (http://www.census.gov/tiger/tms/gazetteer/county2k.txt). United States Census. .
Retrieved 2011-02-13.

[5] http://www.census.gov/population/www/censusdata/cencounts/files/de190090.txt

[6] http://factfinder2.census.gov

[7] http://mapserver.lib.virginia.edu/

[8] "American FactFinder" (http://factfinder.census.gov). United States Census Bureau. . Retrieved 2008-01-31.

External links

- Kent County webpage (http://www.co.kent.de.us/)
- Kent County & Greater Dover Convention and Visitors Bureau (http://www.visitdover.com/)

Delaware_Route_300

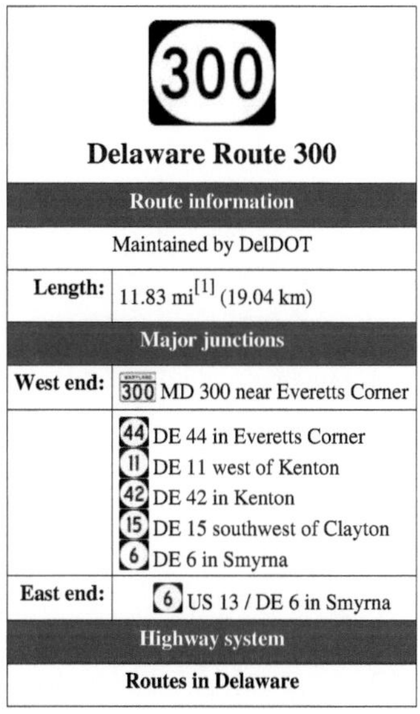

Delaware Route 300

Route information	
Maintained by DelDOT	
Length:	11.83 mi[1] (19.04 km)
Major junctions	
West end:	MD 300 near Everetts Corner
	DE 44 in Everetts Corner DE 11 west of Kenton DE 42 in Kenton DE 15 southwest of Clayton DE 6 in Smyrna
East end:	US 13 / DE 6 in Smyrna
Highway system	
Routes in Delaware	

Delaware Route 300 is a state highway in Kent County, Delaware. The route is a continuation of Maryland Route 300 from the Maryland border near Everetts Corner. It runs in a northeast direction from there to Smyrna, where it ends at U.S. Route 13. The road was first built as a state highway in the 1920s and 1930s between the Maryland border and Clayton, with the DE 300 designation given to the road by 1936. The route was extended to its current terminus in the 1950s.

Route description

DE 6/DE 300 eastbound approaching US 13 in Smyrna

Delaware Route 300 heads east from the Maryland border on Sudlersville Road. Just across the state line, the route intersects the western terminus of Delaware Route 44 in Everetts Corner. It then heads east through farmland and intersects the northern terminus of Delaware Route 11. It then heads into the town of Kenton, where it is known as Main Street. It intersects Delaware Route 42 within Kenton. Past Kenton, DE 300 heads to the northeast on Wheatleys Pond Road. Midway between Kenton and Clayton, it features a brief concurrency with Delaware Route 15. DE 300 then heads north toward Clayton, passing to the south of the center of the town. Between Clayton and Smyrna, Delaware Route 6 merges onto DE 300, and the two routes head northeast through the northern part of Smyrna on Glenwood Avenue. The road ends at U.S. Route 13, where DE 300 finds its eastern terminus and DE 6 makes a right turn to overlap with US 13.[1] [2]

History

By 1920, what is now DE 300 existed as an unimproved county road.[3] The road was completed as a state highway between Kenton and Clayton and was proposed as one west of Kenton by 1924.[4] By 1931, the entire route between Clayton and the Maryland border was completed as a state highway.[5] When Delaware created its state highway system by 1936, DE 300 was routed on its current alignment between the Maryland border and an intersection with DE 6 between Clayton and Smyrna.[6] By 1954, DE 300 was extended east to its present terminus at US 13 in Smyrna.[7]

Major intersections

The entire route is in Kent County.

Location	Mile[1]	Road	Notes
	0.00	300 MD 300 west (Sudlersville Road)	Maryland state line, western terminus
Everetts Corner	0.33	44 DE 44 east (Everetts Corner Road)	
	4.54	11 DE 11 south (Arthursville Road)	
Kenton	5.32	42 DE 42 (Commerce Street)	
		15 DE 15 south (Mt. Friendship Road)	West end of DE 15 overlap
		15 DE 15 north (Alley Corner Road)	East end of DE 15 overlap
Smyrna	10.73	6 DE 6 west (Main Street)	West end of DE 6 overlap
	11.83	6 US 13 / DE 6 east (Dupont Boulevard)	Eastern terminus
1.000 mi = 1.609 km; 1.000 km = 0.621 mi			

References

[1] http://www.deldot.gov/information/pubs_forms/manuals/traffic_counts/2006/pdf/rpt_pgs1_38_rev.pdf DelDOT 2006 Traffic Count and Mileage Report

[2] Google, Inc. *Google Maps – overview of Delaware Route 300* (http://maps.google.com/maps?f=d&source=s_d&saddr=delaware+300+and+maryland+300&daddr=us+13+and+delaware+300&hl=en&geocode=FdMiVgIdmhx8-ym_Ge7SNoLHiTFrpGXqP-ed8A;FRrDVwIdfFR--ykV-EtSOHHHiTHsRzUFWUOgTA&mra=ls&sll=39.293923,-75.614433&sspn=0.114251,0.338173&ie=UTF8&ll=39.252727,-75.679665&spn=0.114318,0.338173&t=h&z=12) (Map). Cartography by Google, Inc. . Retrieved 2010-08-22.

[3] Delaware Department of Transportation (PDF). *Delaware Official Highway Map* (http://www.deldot.gov/archaeology/historic_pres/historic_highway_maps/pdf/cd_002.pdf) (Map) (1920 ed.). . Retrieved 2010-04-15.

[4] Delaware Department of Transportation (PDF). *Delaware Official Highway Map* (http://www.deldot.gov/archaeology/historic_pres/historic_highway_maps/pdf/cd_003.pdf) (Map) (1924 ed.). . Retrieved 2010-08-01.

[5] Delaware Department of Transportation (PDF). *Delaware Official Highway Map* (http://www.deldot.gov/archaeology/historic_pres/historic_highway_maps/pdf/cd_005.pdf) (Map) (1931 ed.). . Retrieved 2010-08-19.

[6] Delaware Department of Transportation (PDF). *Delaware Official Highway Map* (http://www.deldot.gov/archaeology/historic_pres/historic_highway_maps/pdf/cd_008.pdf) (Map) (1936/37 ed.). . Retrieved 2010-04-15.

[7] Delaware Department of Transportation (PDF). *Delaware Official Highway Map* (http://www.deldot.gov/archaeology/historic_pres/historic_highway_maps/pdf/cd_028.pdf) (Map) (1954/55 ed.). . Retrieved 2010-08-22.

Delaware_Route_8

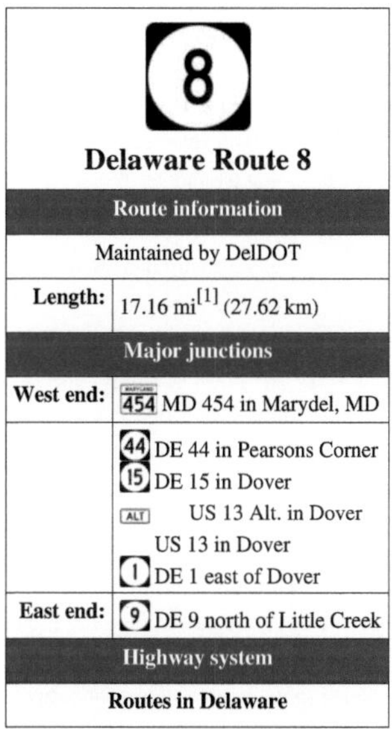

Delaware Route 8 is a state highway located in Kent County, Delaware. It runs from Maryland Route 454 at the Maryland border near Marydel, Maryland east to an intersection with Delaware Route 9 north of Little Creek. The route passes through the heart of Delaware's capital, Dover, on Division Street and the western part of Dover on Forrest Avenue. The road was built as a state highway west of Dover by 1924 and east of Dover by 1931. The DE 8 designation was given to the road by 1936.

Route description

Delaware Route 8 heads to the northeast from the Maryland border in the stateline town of Marydel on Halltown Road. In Pearsons Corner, it intersects the eastern terminus of Delaware Route 44, where the name changes to Forrest Avenue. It then heads east through the countryside west of Dover, which is home to many Amish families.[1][2]

Westbound Delaware Route 8 (Forrest Avenue) in West Dover.

DE 8 then crosses into Dover, where it widens from a two-lane country road to a five-lane suburban road. It passes through the residential and commercial areas of West Dover and it intersects Delaware Route 15. It then heads toward downtown Dover, where DE 8 bears to the left onto Division Street, narrowing to a two-lane road. It passes through the northern edge of downtown Dover, intersecting U.S. Route 13 Alternate (Governors Avenue) and passing by Wesley College.[1] [2]

It then leaves downtown Dover and intersects U.S. Route 13. The name then changes to North Little Creek Road and it continues through the eastern part of Dover. At the eastern edge of Dover, it features a partial interchange with Delaware Route 1, providing access to and from the north. This interchange was not originally constructed when DE 1 was built in the early 1990s; it was built later by upgrading an emergency connection that existed between the two routes. It then heads east through countryside to an intersection with Delaware Route 9 just north of the town of Little Creek.[1] [2]

History

By 1920, what is now DE 8 existed as an unimproved county road.[3] The route was completed as a state highway between the Maryland border in Marydel and Dover by 1924.[4] By 1925, the road was proposed as a state highway between Dover and Little Creek.[5] This state highway was completed by 1931.[6] When Delaware created its state highway system by 1936, DE 8 was assigned to its current alignment between the Maryland border in Marydel and DE 9 north of Little Creek.[7] The route has remained on the same roads since its inception.[8]

Major intersections

The entire route is in Kent County.

Location	Mile[1]	Road	Notes
Marydel	0.00	454 MD 454 north (Crown Stone Road)	Maryland state line, western terminus
Pearsons Corner	6.21	44 DE 44 west (Hartly Road)	
Dover	12.00	15 DE 15 (Saulsbury Road)	
	12.81	ALT US 13 Alt. (Governors Avenue)	
	13.54	US 13 (Dupont Highway)	
		1 DE 1	Interchange
Little Creek	17.16	9 DE 9 (Bayside Drive)	Eastern terminus
1.000 mi = 1.609 km; 1.000 km = 0.621 mi			

References

[1] http://www.deldot.gov/information/pubs_forms/manuals/traffic_counts/2006/pdf/rpt_pgs1_38_rev.pdf DelDOT 2006 Traffic Count and Mileage Report

[2] Google, Inc. *Google Maps – overview of Delaware Route 8* (http://maps.google.com/maps?f=d&source=s_d&saddr=delaware+8+and+maryland+454&daddr=Delaware+9+&+State+Road+8,+Dover,+Kent,+Delaware+19901&geocode=FaLPVAIdkjZ8-ynVhVIyboDHiTHADmVoFi6Wfg;FcS1VQIdkbiA-ym9FfZ1PGTHiTEuJJuzykz91Q&hl=en&mra=pd&mrcr=0&sll=39.142067,-75.597933&sspn=0.114498,0.338173&ie=UTF8&ll=39.141245,-75.597954&spn=0.1145,0.338173&t=h&z=12) (Map). Cartography by Google, Inc. . Retrieved 2010-08-23.

[3] Delaware Department of Transportation (PDF). *Delaware Official Highway Map* (http://www.deldot.gov/archaeology/historic_pres/historic_highway_maps/pdf/cd_002.pdf) (Map) (1920 ed.). . Retrieved 2010-04-15.

[4] Delaware Department of Transportation (PDF). *Delaware Official Highway Map* (http://www.deldot.gov/archaeology/historic_pres/historic_highway_maps/pdf/cd_003.pdf) (Map) (1924 ed.). . Retrieved 2010-08-01.

[5] Delaware Department of Transportation (PDF). *Delaware Official Highway Map* (http://www.deldot.gov/archaeology/historic_pres/historic_highway_maps/pdf/cd_004.pdf) (Map) (1925 ed.). . Retrieved 2010-08-19.

[6] Delaware Department of Transportation (PDF). *Delaware Official Highway Map* (http://www.deldot.gov/archaeology/historic_pres/historic_highway_maps/pdf/cd_005.pdf) (Map) (1931 ed.). . Retrieved 2010-08-19.

[7] Delaware Department of Transportation (PDF). *Delaware Official Highway Map* (http://www.deldot.gov/archaeology/historic_pres/historic_highway_maps/pdf/cd_008.pdf) (Map) (1936/37 ed.). . Retrieved 2010-04-15.

[8] Delaware Department of Transportation (PDF). *Delaware Official Highway Map* (http://www.deldot.gov/archaeology/historic_pres/historic_highway_maps/pdf/cd_083.pdf) (Map) (2008 ed.). . Retrieved 2010-08-02.

Delaware_Route_11

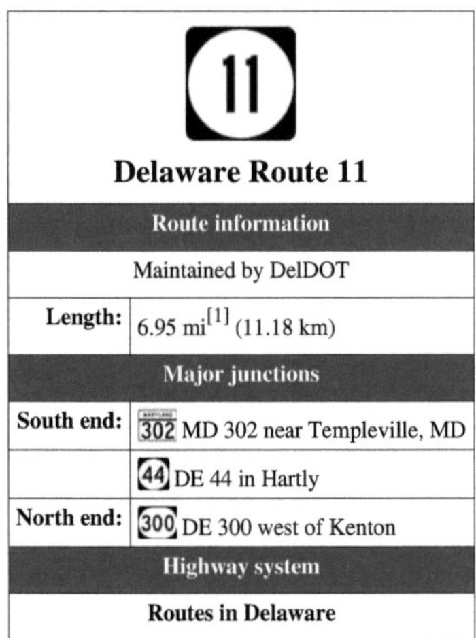

Delaware Route 11

Route information	
Maintained by DelDOT	
Length:	6.95 mi[1] (11.18 km)
Major junctions	
South end:	302 MD 302 near Templeville, MD
	44 DE 44 in Hartly
North end:	300 DE 300 west of Kenton
Highway system	
Routes in Delaware	

Delaware Route 11 is a state highway in Kent County, Delaware. It runs from Maryland Route 302 at the Maryland border near Templeville, Maryland to Delaware Route 300 near Kenton, Delaware. The road, which is known as Arthursville Road its entire length, passes through the farmland of western Kent County. It passes through the town of Hartly, where it intersects Delaware Route 44. DE 11 was created by 1936 on a road taken over by the state by 1931.

Route description

DE 11 begins at the Maryland border in western Kent County, where Maryland Route 302 continues to the west towards Templeville, Maryland. DE 11 heads northeast on two-lane, undivided Arthursville Road through a mix of woods and farms, with occasional residences. The road enters the town of Hartly, where it passes Hartly Elementary School and several homes. In the center of Hartly, the route crosses Delaware Route 44. After DE 44, the road heads north before curving northeast and leaving Hartly. DE 11 continues back into rural areas, making another turn north and passing a few homes. The road runs northeast through more woodland and farm fields before coming to an end at Delaware Route 300 to the southwest of Kenton.[2]

History

By 1920 what is now DE 11 existed as an unimproved county road.[3] [4] The road was taken over by the state and paved by 1931.[5] When Delaware created its state highway system by 1936, DE 11 was assigned to its current alignment.[6] The route has not changed since its inception.[7]

Major intersections

The entire route is in Kent County.

Location	Mile[1]	Road	Notes
	0.00	302 MD 302 west (Barclay Road)	Maryland state line, southern terminus
Hartly	2.58	44 DE 44 (Main Street)	
	6.95	300 DE 300 (Sudlersville Road)	Northern terminus
1.000 mi = 1.609 km; 1.000 km = 0.621 mi			

References

[1] http://www.deldot.gov/information/pubs_forms/manuals/traffic_counts/2006/pdf/rpt_pgs1_38_rev.pdf DelDOT 2006 Traffic Count and Mileage Report

[2] Google, Inc. *Google Maps – overview of Delaware Route 11* (http://maps.google.com/maps?f=d&source=s_d&saddr=Maryland+302+ &+Delaware+11,+Marydel,+Kent&daddr=Delaware+11+and+Delaware+300&hl=en& geocode=FZ1HVQIdfC58-yn1Qge6soHHiTGooTb4RxDXUg;FYZ6VgIdC0h9-ymJGnLgUHjHiTEM8xagZVYmsA&mra=pe&mrcr=0& sll=39.144441,-75.745471&sspn=0.013213,0.0421&ie=UTF8&t=h&z=12) (Map). Cartography by Google, Inc. . Retrieved 2010-03-17.

[3] Delaware Department of Transportation (PDF). *Delaware Official Highway Map* (http://www.deldot.gov/archaeology/historic_pres/ historic_highway_maps/pdf/cd_002.pdf) (Map) (1920 ed.). . Retrieved 2010-04-15.

[4] Delaware Department of Transportation (PDF). *Delaware Official Highway Map* (http://www.deldot.gov/archaeology/historic_pres/ historic_highway_maps/pdf/cd_003.pdf) (Map) (1924 ed.). . Retrieved 2010-08-01.

[5] Delaware Department of Transportation (PDF). *Delaware Official Highway Map* (http://www.deldot.gov/archaeology/historic_pres/ historic_highway_maps/pdf/cd_005.pdf) (Map) (1931 ed.). . Retrieved 2010-08-02.

[6] Delaware Department of Transportation (PDF). *Delaware Official Highway Map* (http://www.deldot.gov/archaeology/historic_pres/ historic_highway_maps/pdf/cd_008.pdf) (Map) (1936/37 ed.). . Retrieved 2010-04-15.

[7] Delaware Department of Transportation (PDF). *Delaware Official Highway Map* (http://www.deldot.gov/archaeology/historic_pres/ historic_highway_maps/pdf/cd_083.pdf) (Map) (2008 ed.). . Retrieved 2010-08-02.

Hartly,_Delaware

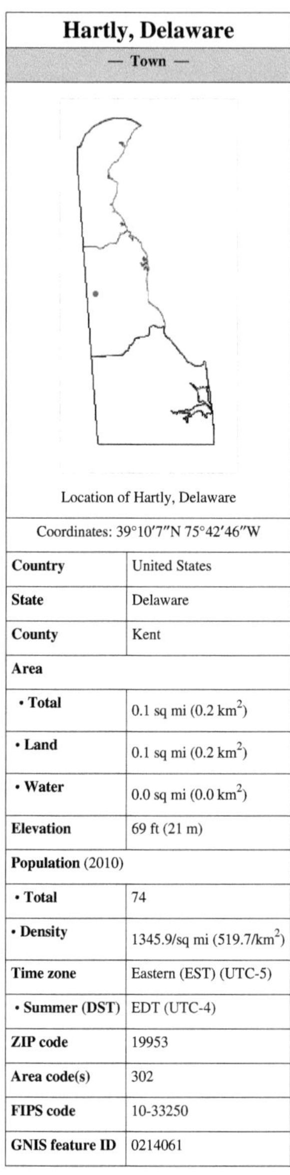

Hartly, Delaware	
— Town —	

Location of Hartly, Delaware

Coordinates: 39°10′7″N 75°42′46″W

Country	United States
State	Delaware
County	Kent
Area	
• Total	0.1 sq mi (0.2 km^2)
• Land	0.1 sq mi (0.2 km^2)
• Water	0.0 sq mi (0.0 km^2)
Elevation	69 ft (21 m)
Population (2010)	
• Total	74
• Density	1345.9/sq mi (519.7/km^2)
Time zone	Eastern (EST) (UTC-5)
• Summer (DST)	EDT (UTC-4)
ZIP code	19953
Area code(s)	302
FIPS code	10-33250
GNIS feature ID	0214061

Hartly is a town in Kent County, Delaware, United States. It is part of the Dover, Delaware Metropolitan Statistical Area. The population was 74 at the 2010 census, making it the least populous municipality in Delaware.[1]

Geography

Hartly is located at 39°10′07″N 75°42′46″W.[2]

According to the United States Census Bureau, the town has a total area of 0.1 square miles (0.26 km^2), all of it land.

Demographics

As of the census[3] of 2000, there were 78 people, 25 households, and 21 families residing in the town. The population density was 1,345.9 people per square mile (501.9/km²). There were 31 housing units at an average density of 534.9 per square mile (199.5/km²). The racial makeup of the town was 91.03% White, 3.85% African American, 1.28% from other races, and 3.85% from two or more races. Hispanic or Latino of any race were 2.56% of the population.

There were 25 households out of which 60.0% had children under the age of 18 living with them, 52.0% were married couples living together, 24.0% had a female householder with no husband present, and 16.0% were non-families. 4.0% of all households were made up of individuals and none had someone living alone who was 65 years of age or older. The average household size was 3.12 and the average family size was 3.29.

In the town the population was spread out with 34.6% under the age of 18, 9.0% from 18 to 24, 39.7% from 25 to 44, 14.1% from 45 to 64, and 2.6% who were 65 years of age or older. The median age was 28 years. For every 100 females there were 81.4 males. For every 100 females age 18 and over, there were 59.4 males.

The median income for a household in the town was $29,375, and the median income for a family was $29,375. Males had a median income of $21,667 versus $24,167 for females. The per capita income for the town was $11,516. There were 14.3% of families and 19.5% of the population living below the poverty line, including 19.2% of under eighteens and none of those over 64.

References

[1] http://www.stateplanning.delaware.gov/census_data_center/

[2] "US Gazetteer files: 2010, 2000, and 1990" (http://www.census.gov/geo/www/gazetteer/gazette.html). United States Census Bureau. 2011-02-12. . Retrieved 2011-04-23.

[3] "American FactFinder" (http://factfinder.census.gov). United States Census Bureau. . Retrieved 2008-01-31.

Delaware_Department_of_Transportation

<table>
<tr><td colspan="2" align="center">Delaware Department of Transportation
(DelDOT)</td></tr>
<tr><td colspan="2" align="center">DelDOT</td></tr>
<tr><td colspan="2" align="center">Agency overview</td></tr>
<tr><td>Preceding agency</td><td>Delaware Highway Department</td></tr>
<tr><td>Jurisdiction</td><td>Delaware</td></tr>
<tr><td>Agency executive</td><td>Shailen Bhatt, Secretary of Transportation</td></tr>
<tr><td>Parent agency</td><td>State of Delaware</td></tr>
<tr><td colspan="2" align="center">Website</td></tr>
<tr><td colspan="2" align="center">http://www.deldot.gov/</td></tr>
</table>

The **Delaware Department of Transportation (DelDOT)** is an agency of the U.S. state of Delaware. The Secretary of Transportation is Shailen Bhatt.[1] The agency has its headquarters in Dover.[2]

The department's responsibilities include maintaining about 90 percent of the state's public roadways, overseeing the "Adopt-A-Highway" program, snow removal, the Division of Motor Vehicles and the Delaware Transit Corporation (known as DART First State).

The DART public transit system was named "Most Outstanding Public Transportation System" in 2003 by the American Public Transportation Association.[3]

DelDOT maintains a 24/7 Traffic Management Center in Smyrna, Delaware at the State Emergency Operations Center. At that location, they monitor traffic conditions, operate traffic lights, and broadcast on 1380 AM via WTMC radio.

Since 1969, the agency has also maintained a transportation library on Bay Road in Dover.

On February 18, 2011, Sec. Carolann Wicks, who had been Secretary of Transportation since 2006, resigned.[4] On March 21, 2011, Cleon Cauley, who had been appointed Deputy Secretary two months earlier, was appointed Acting Secretary.[5] [6] On July 5, 2011, Shailen Bhatt was sworn in as the new Secretary of Transportation.[1]

See also

- Vehicle registration plates of Delaware

References

[1] "Department of Transportation Secretary Bhatt Sworn-In by Governor Markell" (http://www.deldot.gov/public.
    ejs?command=PublicNewsDisplay&id=4030&month=7&year=2011). Delaware Department of Transportation. July 5, 2011. . Retrieved
    July 8, 2011.

[2] " Contact Information (http://www.deldot.gov/home/contact_info/facilities.shtml)." Delaware Department of Transportation. Retrieved
    on October 25, 2009.

[3] "DE Transit - DART First State Garners Prestigious 2003 APTA National Public Transportation System Outstanding Achievement Award"
    (http://www.deldot.gov/public.ejs?command=PublicNewsDisplay&id=1456&month=7&year=2003). *State of Delaware*. July 9, 2003. .
    Retrieved May 11, 2011.

[4] "Governor Accepts Resignation of Transportation Secretary Wicks" (http://governor.delaware.gov/news/2011/1102february/
    20110218-wicks.shtm). February 18, 2011. . Retrieved May 11, 2011.

[5] Kelli Steele (March 17, 2011). "Cleon Cauley Appointed Acting DelDOT Secretary" (http://www.wgmd.com/?p=19985). *WGMD.Com*. .
    Retrieved May 11, 2011.

[6] Chad Livengood (December 2, 2010). "DelDOT gets new deputy secretary" (http://blogs.delawareonline.com/dialoguedelaware/2010/12/
    02/deldot-gets-new-deputy-secretary/). *Delaware Online (Gannett)*. . Retrieved May 11, 2011.

External links

- Official website (http://www.deldot.gov/index.shtml)
- DART First State (http://www.dartfirststate.com/)
- Transportation Library (http://ntl.bts.gov/tldir/libraries/library028.html)

Delaware

<table>
<tr><td colspan="2" align="center">State of Delaware</td></tr>
<tr><td colspan="2" align="center"> 
Flag   Seal</td></tr>
<tr><td colspan="2" align="center">Nickname(s): The First State; The Small Wonder;
Blue Hen State; The Diamond State</td></tr>
<tr><td colspan="2" align="center">Motto(s): Liberty and Independence</td></tr>
<tr><td colspan="2" align="center">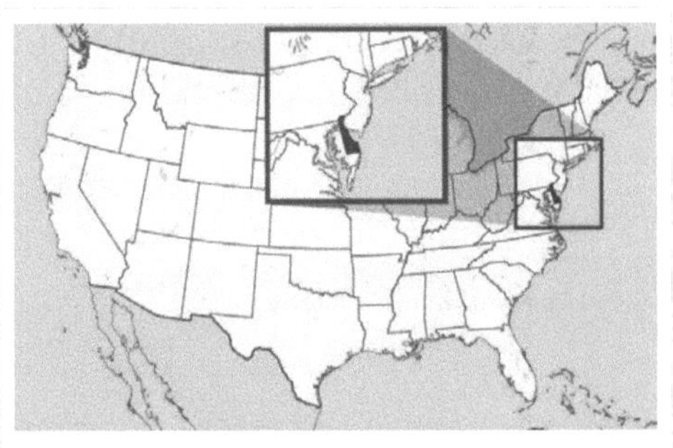</td></tr>
<tr><td>Demonym</td><td>Delawarean</td></tr>
<tr><td>Capital</td><td>Dover</td></tr>
<tr><td>Largest city</td><td>Wilmington</td></tr>
<tr><td>Area</td><td>Ranked 49th in the U.S.</td></tr>
<tr><td>- Total</td><td>2,490 sq mi
(6,452 km^2)</td></tr>
<tr><td>- Width</td><td>30 miles (48 km)</td></tr>
<tr><td>- Length</td><td>96 miles (154 km)</td></tr>
<tr><td>- % water</td><td>21.5</td></tr>
<tr><td>- Latitude</td><td>38° 27′ N to 39° 50′ N</td></tr>
<tr><td>- Longitude</td><td>75° 3′ W to 75° 47′ W</td></tr>
<tr><td>Population</td><td>Ranked 45th in the U.S.</td></tr>
<tr><td>- Total</td><td>897,934[1]</td></tr>
<tr><td>- Density</td><td>442.6/sq mi (170.87/km^2)
Ranked 6th in the U.S.</td></tr>
<tr><td>- Median income</td><td>$50,152 (12th)</td></tr>
<tr><td>Elevation</td><td></td></tr>
</table>

- Highest point	Near the Ebright Azimuth[2] [3] [4] 447 ft (136.2 m)
- Mean	60 ft (20 m)
- Lowest point	Atlantic Ocean[2] sea level
Admission to Union	December 7, 1787 (1st)
Governor	Jack A. Markell (D)
Lieutenant Governor	Matthew P. Denn (D)
Legislature	General Assembly
- Upper house	Senate
- Lower house	House of Representatives
U.S. Senators	Thomas R. Carper (D) Chris Coons (D)
U.S. House delegation	John C. Carney, Jr. (D) (list)
Time zone	Eastern: UTC-5/-4
Abbreviations	DE Del. US-DE
Website	[delaware.gov delaware.gov]

Delaware (◀ i/ˈdɛləwɛər/ *del-ə-wair*)[5] is a U.S. state located on the Atlantic Coast in the Mid-Atlantic region of the United States. It is bordered to the south and west by Maryland, and to the north by Pennsylvania.[6] The state takes its name from Thomas West, 3rd Baron De La Warr, an English nobleman and Virginia's first colonial governor, after whom what is now called Cape Henlopen was originally named.[7]

Delaware is located in the northeastern portion of the Delmarva Peninsula and is the second smallest state in area (after Rhode Island). Estimates in 2007 rank the population of Delaware as 45th in the nation, but 6th in population density, with more than 60% of the population in New Castle County.[8] Delaware is divided into three counties. From north to south, these three counties are New Castle, Kent, and Sussex. While the southern two counties have historically been predominantly agricultural, New Castle County has been more industrialized.

The state ranks second in civilian scientists and engineers as a percentage of the workforce and number of patents issued to companies or individuals per 1,000 workers.[9] The history of the state's economic and industrial development is closely tied to the impact of the Du Pont family, founders and scions of E. I. du Pont de Nemours and Company, one of the world's largest chemical companies.[10]

Before its coastline was first explored by Europeans in the 16th century, Delaware was inhabited by several groups of Native Americans, including the Lenape in the north and Nanticoke in the south. It was initially colonized by Dutch traders at Zwaanendael, located near the present town of Lewes, in 1631.[11] Delaware was one of the 13 colonies participating in the American Revolution and on December 7, 1787, became the first state to ratify the Constitution of the United States, thereby becoming known as *The First State.*

Geography

Delaware is 96 miles (154 km) long and ranges from 9 miles (14 km) to 35 miles (56 km) across, totaling 1954 square miles (5060 km^2), making it the second-smallest state in the United States after Rhode Island. Delaware is bounded to the north by Pennsylvania; to the east by the Delaware River, Delaware Bay, New Jersey and the Atlantic Ocean; and to the west and south by Maryland. Small portions of Delaware are also situated on the eastern side of the Delaware River sharing land boundaries with New Jersey. The state of Delaware, together with the Eastern Shore counties of Maryland and two counties of Virginia, form the Delmarva Peninsula, which stretches down the Mid-Atlantic Coast.

The definition of the northern boundary of the state is unusual. Most of the boundary between Delaware and Pennsylvania was originally defined by an arc extending 12 miles (19.3 km) from the cupola of the courthouse in the city of New Castle. This boundary is often referred to as the Twelve-Mile Circle.[12] This is the only nominally circular state boundary in the United States.

This border extends all the way east to the low-tide mark on the New Jersey shore, then continues south along the shoreline until it again reaches the 12-mile (19 km) arc in the south; then the boundary continues in a more conventional way in the middle of the main channel (thalweg) of the Delaware River. To the west, a portion of the arc extends past the easternmost edge of Maryland. The remaining western border runs slightly east of due south from its intersection with the arc. The Wedge of land between the northwest part of the arc and the Maryland border was claimed by both Delaware and Pennsylvania until 1921, when Delaware's claim was confirmed.

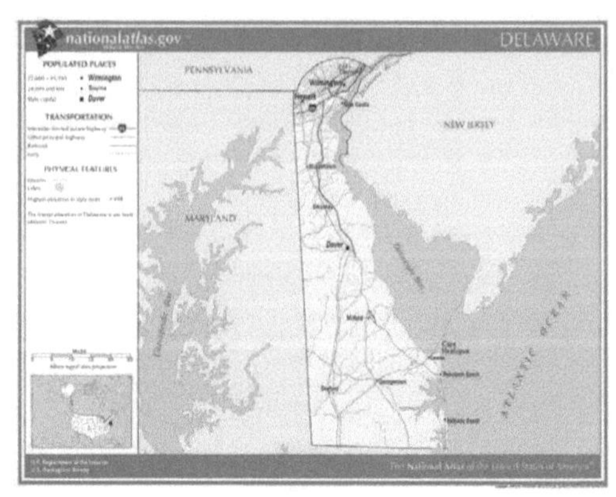

Map of Delaware

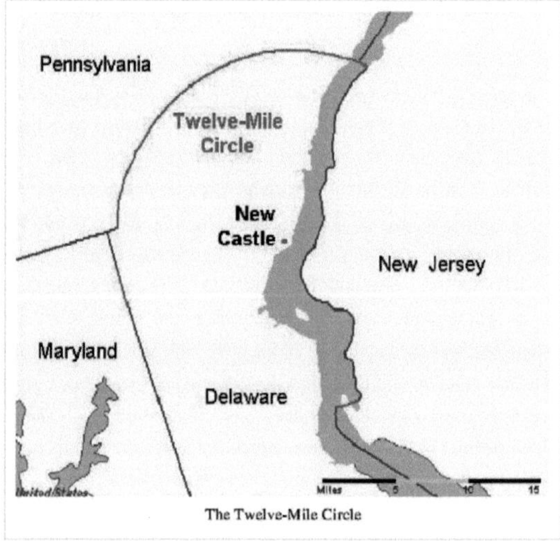

The Twelve-Mile Circle

Topography

Delaware is on a level plain, with the lowest mean elevation of any state in the nation.[2] Its highest elevation, located at Ebright Azimuth, near Concord High School, does not rise fully 450 feet (140 m) above sea level.[2] The northernmost part of the state is part of the Appalachian Piedmont with hills and rolling surfaces. The fall line approximately follows the Robert Kirkwood Highway between Newark and Wilmington; south of this road is the Atlantic Coastal Plain with flat, sandy, and, in some parts, swampy ground. A ridge about 75 to 80 feet (23 to 24 m) in elevation extends along the western boundary of the state and separates between the watersheds that feed Delaware River and Bay to the east and the Chesapeake Bay to the west.

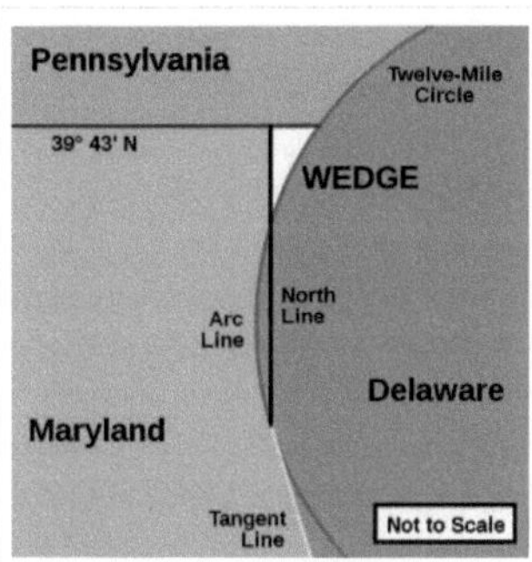

Diagram of the Twelve-Mile Circle, the Mason-Dixon Line and "The Wedge". All blue and white areas are inside Delaware.

Climate

Since almost all of Delaware is a part of the Atlantic Coastal Plain, the effects of the ocean moderate its climate. The state is in a transitional zone between a humid subtropical climate and a continental climate. Despite its small size (roughly 100 miles (160 km) from its northernmost to southernmost points), there is significant variation in mean temperature and amount of snowfall between Sussex County and New Castle County. The southern portion of the state has a somewhat milder climate and a longer growing season than the northern portion of the state. Delaware's all time record high of 110 °F (43 °C) was recorded at Millsboro on July 21, 1930; the all time record low of −17 °F (−27 °C) was also recorded at Millsboro on January 17, 1893.

Environment

The transitional climate of Delaware supports a wide variety of vegetation. In the northern third of the state are found Northeastern coastal forests and mixed oak forests typical of the northeastern United States.[13] In the southern two-thirds of the state are found Middle Atlantic coastal forests.[13] Trap Pond State Park in Sussex County, for example, supports what may be one of the northernmost stands of bald cypress.

Environmental management

Delaware provides government subsidy support for the clean-up of property "lightly contaminated" by hazardous waste, the proceeds for which come from a tax on wholesale petroleum sales.[14]

History

Native Americans

Before Delaware was settled by European colonists, the area was home to the Eastern Algonquian tribes known as the Unami Lenape or Delaware throughout the Delaware valley, and the Nanticoke along the rivers leading into the Chesapeake Bay. The Unami Lenape in the Delaware Valley were closely related to Munsee Lenape tribes along the Hudson River. They had a settled hunting and agricultural society, and they rapidly became middlemen in an increasingly frantic fur trade with their ancient enemy, the Minqua or Susquehannock. With the loss of their lands on the Delaware River and the destruction of the Minqua by the Iroquois of the Five Nations in the 1670s, the remnants of the Lenape who wished to remain identified as such left the region and moved over the Alleghany Mountains by the mid-18th century. Generally, those who did not relocate out of the State of Delaware were baptized, became Christian and were grouped together with other persons of color in official records and in the minds of their non-Native American neighbors.

Colonial Delaware

The Dutch were the first Europeans to settle in present-day Delaware in the Middle region by establishing a trading post at Zwaanendael, near the site of Lewes in 1631. Within a year all the settlers were killed in a dispute with area Indian tribes. In 1638 New Sweden, a Swedish trading post and colony, was established at Fort Christina (now in Wilmington) by Peter Minuit at the head of a group of Swedes, Finns and Dutch. The colony of New Sweden lasted for 17 years. In 1651, the Dutch, reinvigorated by the leadership of Peter Stuyvesant, established a fort at present-day New Castle, and in 1655 they conquered the New Sweden colony, annexing it into the Dutch New Netherland.[15] Only nine years later, in 1664, the Dutch were conquered by a fleet of English ships by Sir Robert Carr under the direction of James, the Duke of York. Fighting off a prior claim by Cecilius Calvert, 2nd Baron Baltimore, Proprietor of Maryland, the Duke passed his somewhat dubious ownership on to William Penn in 1682. Penn strongly desired access to the sea for his Pennsylvania province and leased what then came to be known as the "Lower Counties on the Delaware" [15] from the Duke.

Penn established representative government and briefly combined his two possessions under one General Assembly in 1682. However, by 1704 the Province of Pennsylvania had grown so large that their representatives wanted to make decisions without the assent of the Lower Counties and the two groups of representatives began meeting on their own, one at Philadelphia, and the other at New Castle. Penn and his heirs remained proprietors of both and always appointed the same person Governor for their Province of Pennsylvania and their territory of the Lower Counties. The fact that Delaware and Pennsylvania shared the same governor was not unique. From 1703-1738, New York and New Jersey shared a governor.[16] Massachusetts and New Hampshire also shared a governor for some time.[17]

Dependent in early years on indentured labor, Delaware imported more slaves as the number of English immigrants decreased with better economic conditions in England. The colony became a slave society and cultivated tobacco as a cash crop, although English immigrants continued to arrive.

American Revolution

Like the other middle colonies, the Lower Counties on the Delaware initially showed little enthusiasm for a break with Britain. The citizenry had a good relationship with the Proprietary government, and generally were allowed more independence of action in their Colonial Assembly than in other colonies. Merchants at the port of Wilmington had trading ties with the British.

So it was that New Castle lawyer Thomas McKean denounced the Stamp Act in the strongest terms, and Kent County native John Dickinson became the "Penman of the Revolution." Anticipating the Declaration of Independence, Patriot leaders Thomas McKean and Caesar Rodney convinced the Colonial Assembly to declare itself separated from British and Pennsylvania rule on June 15, 1776. The person best representing Delaware's majority, George Read, could not bring himself to vote for a Declaration of Independence. Only the dramatic overnight ride of Caesar Rodney gave the delegation the votes needed to cast Delaware's vote for independence.

Initially led by John Haslet, Delaware provided one of the premier regiments in the Continental Army, known as the "Delaware Blues" and nicknamed the "Blue Hen Chickens." In August 1777, General Sir William Howe led a British army through Delaware on his way to a victory at the Battle of Brandywine and capture of the city of Philadelphia. The only real engagement on Delaware soil was the Battle of Cooch's Bridge, fought on September 3, 1777, at Cooch's Bridge in New Castle County.

Following the Battle of Brandywine, Wilmington was occupied by the British, and State President John McKinly was taken prisoner. The British remained in control of the Delaware River for much of the rest of the war, disrupting commerce and providing encouragement to an active Loyalist portion of the population, particularly in Sussex County. Because the British promised slaves of rebels freedom for fighting with them, escaped slaves flocked north to join their lines.[18]

Following the American Revolution, statesmen from Delaware were among the leading proponents of a strong central United States with equal representation for each state.

Slavery and race

Many colonial settlers came to Delaware from Maryland and Virginia, which had been experiencing a population boom. The economies of these colonies were chiefly based on tobacco culture and were increasingly dependent on slave labor for its intensive cultivation. Most of the English colonists arrived as indentured servants, hiring themselves out as laborers for a fixed period to pay for their passage. In the early years the line between indentured servants and African slaves or laborers was fluid. Most of the free African-American families in Delaware before the Revolution had migrated from Maryland to find more affordable land. They were descendants chiefly of relationships or marriages between servant women and enslaved, servant or free African or African-American men.[19] As the flow of indentured laborers to the colony decreased with improving economic conditions in England, more slaves were imported for labor.

At the end of the colonial period, the number of enslaved people in Delaware began to decline. Shifts in the agriculture economy from tobacco to mixed farming created less need for slaves' labor. Local Methodists and Quakers encouraged slaveholders to free their slaves following the American Revolution, and many did so in a surge of individual manumissions for idealistic reasons. By 1810 three-quarters of all blacks in Delaware were free. When John Dickinson freed his slaves in 1777, he was Delaware's largest slave owner with 37 slaves. By 1860 the largest slaveholder owned only 16 slaves.[20]

Although attempts to abolish slavery failed by narrow margins in the legislature, in practical terms, the state had mostly ended the practice. By the 1860 census on the verge of the Civil War, 91.7 percent of the black population, or nearly 20,000 people, were free.[21] [22]

The independent black denomination was chartered by freed slave Peter Spencer in 1813 as the "Union Church of Africans". This followed the 1793 establishment of the African Methodist Episcopal Church in Philadelphia, which had ties to the Methodist Episcopal Church until 1816. Spencer built a church in Wilmington for the new denomination.[23] This was renamed the African Union First Colored Methodist Protestant Church and Connection, more commonly known as the A.U.M.P. Church. Begun by Spencer in 1814, the annual gathering of the Big August Quarterly still draws people together in a religious and cultural festival, the oldest such cultural festival in the nation.

At the onset of the Civil War, Delaware was only nominally a slave state, and it remained in the Union. Delaware voted against secession on January 3, 1861. As the governor said, Delaware had been the first state to embrace the Union by ratifying the Constitution and would be the last to leave it. While most Delaware citizens who fought in the war served in the regiments of the state, some served in companies on the Confederate side in Maryland and Virginia Regiments. Delaware is notable for being the only slave state from which no Confederate regiments or militia groups were assembled. It freed the remaining 1,000 or so Delaware slaves with the ratification of the 13th Amendment to the US Constitution in December 1865.

Demographics

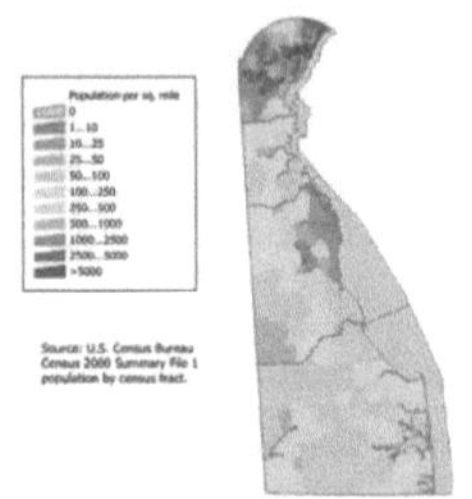

Delaware Population Density Map

Historical populations		
Census	Pop.	%±
1790	59096	—
1800	64273	8.8%
1810	72674	13.1%
1820	72749	0.1%
1830	76748	5.5%
1840	78085	1.7%
1850	91532	17.2%
1860	112216	22.6%
1870	125015	11.4%
1880	146608	17.3%
1890	168493	14.9%
1900	184735	9.6%
1910	202322	9.5%
1920	223003	10.2%
1930	238380	6.9%
1940	266505	11.8%
1950	318085	19.4%
1960	446292	40.3%
1970	548104	22.8%
1980	594338	8.4%
1990	666168	12.1%
2000	783600	17.6%
2010	897934	14.6%
Sources: 1910-2010[24]		

According to the 2010 U.S. Census, Delaware had a population of 897,934. In terms of race and ethnicity, the state was 68.9% White (65.3% Non-Hispanic White Alone), 21.4% Black or African American, 0.5% American Indian and Alaska Native, 3.2% Asian, 0.0% Native Hawaiian and Other Pacific Islander, 3.4% from Some Other Race, and 2.7% from Two or More Races. Hispanics and Latinos of any race made up 8.2% of the population.[25]

Delaware is the sixth most densely populated state, with a population density of 442.6 people per square mile, 356.4 per square mile more than the national average, and ranking 45th in population. Only the states of Delaware, West Virginia, Vermont, Maine, North Dakota, and Wyoming do not have a single city with a population over 100,000 as of the 2007 census estimates.[8] The center of population of Delaware is located in New Castle County, in the town of Townsend.[26]

Languages

As of 2000, 91% of Delaware residents age 5 and older speak only English at home; 5% speak Spanish. French is the third most spoken language at 0.7%, followed by Chinese at 0.5% and German at 0.5%.

Legislation had been proposed in both the House and the Senate in Delaware to designate English as the official language.[27] [28] Neither bill was passed in the legislature.

Religion

The religious affiliations of the people of Delaware are:

- Methodist – 20%
- Baptist – 19%
- No Religion – 17%
- Roman Catholic – 9%
- Lutheran – 4%
- Presbyterian – 3%
- Pentecostal – 3%
- Episcopalian/Anglican - 2%
- Seventh-day Adventist - 2%
- Churches of Christ - 1%
- Other Christian – 3%
- Muslim - 2%
- Jewish - 1%
- Other – 5%
- Refused - 9%

(Source: American Religious Identification Survey, City University of New York [29])

As of the year 2000, The Association of Religion Data Archives[30] reported that the three largest denominational groups in Delaware are Catholic, Mainline Protestant, and Evangelical Protestant. The Catholic Church has the highest number of adherents in Delaware (at 151,740), followed by the United Methodist Church with 59,471 members reported and the Presbyterian Church (U.S.A.), reporting 14,880 adherents. The religious body with the largest number of congregations is the United Methodist Church (with 162 congregations) followed by the Catholic Church (with 46 congregations).

The Roman Catholic Diocese of Wilmington and the Episcopal Diocese of Delaware oversee the parishes within their denominations. The A.U.M.P. Church, the oldest African-American denomination in the nation, was founded in Wilmington. It still has a substantial presence in the state. Reflecting new immigrant populations, an Islamic mosque has been built in the Ogletown area, and a Hindu temple in Hockessin.

Economy

The gross state product of Delaware in 2010 was $62.3 billion.[31]

Affluence

Average sale price for new & existing homes (in US$)[32]

DE County	March 2010	March 2011
New Castle	229,000	216,000
Sussex	323,000	296,000
Kent	186,000	178,000

The per capita personal income was $34,199, ranking 9th in the nation. In 2005, the average weekly wage was $937, ranking 7th in the nation.[33]

In common with many counties in the United States, each of the three Delaware counties have experienced a year-on-year decreasing in the sales price of new and existing homes when comparing 2010 to 2011.[32]

Agriculture

Delaware's agricultural output consists of poultry, nursery stock, soybeans, dairy products and corn. Its industrial outputs include chemical products, automobiles, processed foods, paper products, and rubber and plastic products. Delaware's economy generally outperforms the national average of the United States.

Industries

As of January 2011, the state's unemployment rate was 8.5%.[34] The state's largest employers are:

"Picking Peaches in Delaware" from an 1878 issue of Harper's Weekly

- government (State of Delaware, New Castle County)
- education (University of Delaware)
- banking (Bank of America, Wilmington Trust, First USA / Bank One / JPMorgan Chase, AIG, Citigroup, Deutsche Bank, Barclays plc)
- chemical and pharmaceutical companies (E.I. du Pont de Nemours & Co.,[35] [36] Syngenta, AstraZeneca,[37] and Hercules, Inc.)
- healthcare (Christiana Care Health System, Alfred I. duPont Hospital for Children)

    As of 2011, there are approximately 2,800 doctors practicing in the state.[38]

- automotive manufacturing (Fisker Automotive)
- farming, specifically chicken farming in Sussex County (Perdue Farms, Mountaire Farms, Allen Family Foods)

The Dover Air Force Base, located next to the state capital of Dover, is one of the largest Air Force bases in the country and is a major employer in Delaware. In addition to its other responsibilities in the USAF Air Mobility Command, this air base serves as the entry point and mortuary for American military personnel, and some U.S. government civilians, who die overseas.

Incorporation in Delaware

Over 50% of US publicly traded corporations and 60% of the Fortune 500 companies are incorporated in Delaware;[39] the state's attractiveness as a corporate haven is largely because of its business-friendly corporation law. Franchise taxes on Delaware corporations supply about one-fifth of its state revenue.[40] Although Delaware is ranked first tax haven in the world by Tax Justice Network,[41] it is not listed on the OECD's 2009 "Black List", despite objections of Luxembourg's and Switzerland's authorities.[42]

Food and drink

Title 4, chapter 7 of the Delaware Code stipulates that alcoholic liquor only be sold in specifically licensed establishments, and only between 9:00 AM and 1:00 AM.[43] Until 2003, Delaware was among the several states enforcing blue laws and banned sale of liquor on Sunday.[44]

Transportation

The transportation system in Delaware is under the governance and supervision of the Delaware Department of Transportation, also known as "DelDOT".[46] [47] DelDOT manages programs such as a Delaware Adopt-a-Highway program, major road route snow removal, traffic control infrastructure (signs and signals), toll road management, Delaware Division of Motor Vehicles, the Delaware Transit Corporation (branded as "DART First State", the state government public transportation organization), among others. Almost 90 percent of the state's public roadway miles are under the direct maintenance of DelDOT which far exceeds the United States national average of 20

The current state license plate design was introduced in 1959, making it the longest-running license plate design in United States history.[45]

percent for state department of transportation maintenance responsibility; the remaining public road miles are under the supervision of individual municipalities.

Roads

Further information: List of numbered routes in Delaware

One major branch of the U.S. Interstate Highway System, Interstate 95, crosses Delaware southwest-to-northeast across New Castle County. In addition to I-95, there are six U.S. highways that serve Delaware: U.S. Route 9, U.S. Route 13, U.S. Route 40, U.S. Route 113, U.S. Route 202, and U.S. Route 301. There are also several state highways that cross the state of Delaware; a few of them include Delaware Route 1, Delaware Route 9, and Delaware Route 404. U.S. 13 and DE Rt 1 are primary north-south highways connecting Wilmington and Pennsylvania with Maryland, with DE 1 serving as the main route between Wilmington and the Delaware beaches. DE Rt. 9 is a north-south highway connecting Dover and Wilmington via a scenic route along the Delaware Bay. U.S. 40, is a primary east-west route,

Delaware Route 1, a partial toll road linking Fenwick Island and Wilmington.

connecting Maryland with New Jersey. DE Rt. 404 is another primary east-west highway connecting the Chesapeake Bay Bridge in Maryland with the Delaware beaches. The state also operates two toll highways, the Delaware Turnpike, which is Interstate 95, between Maryland and New Castle and the Korean War Veterans Memorial Highway, which is DE Rt. 1, between Wilmington and Dover.

A bicycle route, Delaware Bicycle Route 1, spans the north-south length of the state from the Maryland border in Fenwick Island to the Pennsylvania border north of Montchanin. It is the first of several signed bike routes planned in Delaware.[48]

Delaware has around 1,450 bridges, ninety-five percent of which are under the supervision of DelDOT. About 30 percent of all Delaware bridges were built prior to 1950 and about 60 percent of the number are included in the National Bridge Inventory. Some bridges not under DelDOT supervision includes the four bridges on the Chesapeake and Delaware Canal, which are under the jurisdiction of the U.S. Army Corps of Engineers, and the Delaware Memorial Bridge, which is under the bi-state Delaware River and Bay Authority.

It has been noted that the tar and chip composition of secondary roads in Sussex County make them more prone to deterioration than asphalt roadways found in the rest of the state.[49] Among these roads, Sussex (county road) 238 is among the most problematic.[49]

Ferries

There are three ferries that operate in the state of Delaware:

- Cape May-Lewes Ferry crosses the mouth of the Delaware Bay between Lewes, Delaware and Cape May, New Jersey.
- Woodland Ferry is a cable ferry that crosses the Nanticoke River southwest of Seaford.
- Three Forts Ferry Crossing connects Delaware City with Fort Delaware and Fort Mott in New Jersey

Rail and bus

A Norfolk Southern locomotive in Dover.

Amtrak has two stations in Delaware along the Northeast Corridor; the relatively quiet Newark Rail Station in Newark, and the busier Wilmington Rail Station in Wilmington. The Northeast Corridor is also served by SEPTA's Wilmington/Newark Line of Regional Rail, which serves Claymont, Wilmington, Churchmans Crossing, and Newark. The major freight railroad in Delaware is the Class 1 Norfolk Southern, which provides service to most of Delaware. It connects with two shortline railroads, the Delaware Coast Line Railway and the Maryland & Delaware Railroad. These two shortlines serve local customers in Sussex County. Another Class 1 railroad, CSX, passes through northern New Castle County parallel to the Amtrak Northeast Corridor.

The public transportation system, DART First State, was named "Most Outstanding Public Transportation System" in 2003 by the American Public Transportation Association. Coverage of the system is broad within northern New Castle County with close association to major highways in Kent and Sussex counties. The system includes bus, subsidized passenger rail operated by Philadelphia transit agency SEPTA, and subsidized taxi and paratransit modes, the latter consisting of a state-wide door-to-door bus service for the elderly and disabled.

Air

Delaware is the only state in the United States without commercial air service. New Castle Airport near Wilmington has been served by commercial airlines in the past, the last being Skybus Airlines, which provided service to Columbus, Ohio and Greensboro, North Carolina from March 7, 2008[50] until its bankruptcy on April 5, 2008.

Delaware is centrally situated in the Northeast Corridor region of cities along I-95. Therefore, Delaware commercial airline passengers most frequently use Philadelphia International Airport (PHL) and Baltimore-Washington International Thurgood Marshall Airport (BWI) for domestic and international transit. Residents of Sussex County will also use Wicomico Regional Airport, as it is located less than 10 miles (16 km) from the Delaware border. Newark Liberty International Airport (EWR) and Ronald Reagan Washington National Airport (DCA) are also within a 100-mile (160 km) radius of New Castle County.

The large Dover Air Force Base of the USAF Air Mobility Command is located in the central part of the state, and it is the home of the 436th Airlift Wing and the 512th Airlift Wing.

Other general aviation airports in Delaware include Summit Airport near Middletown, Delaware Airpark near Cheswold, and Sussex County Airport near Georgetown.

Law and government

Delaware's fourth and current constitution, adopted in 1897, provides for executive, judicial and legislative branches.

Legislative branch

The Delaware General Assembly consists of a House of Representatives with 41 members and a Senate with 21 members. It sits in Dover, the state capital. Representatives are elected to two-year terms, while senators are elected to four-year terms. The Senate confirms judicial and other nominees appointed by the governor.

Delaware's U.S. Senators are Thomas R. Carper (Democrat) and Chris Coons (Democrat). Delaware's single U.S. Representative is John Carney (Democrat).

The Delaware General Assembly meet in the Legislative Hall in Dover.

Judicial branch

The Delaware Constitution establishes a number of courts:

- The Delaware Supreme Court is the state's highest court.
- The Delaware Superior Court is the state's trial court of general jurisdiction.
- The Delaware Court of Chancery deals primarily in corporate disputes.
- The Family Court handles domestic and custody matters.
- The Delaware Court of Common Pleas has jurisdiction over a limited class of civil and criminal matters.

Minor non-constitutional courts include the Justice of the Peace Courts and Aldermen's Courts.

Significantly, Delaware has one of the few remaining Courts of Chancery in the nation, which has jurisdiction over equity cases, the vast majority of which are corporate disputes, many relating to mergers and acquisitions. The Court of Chancery and the Supreme Court have developed a worldwide reputation for rendering concise opinions concerning corporate law which generally (but not always) grant broad discretion to corporate boards of directors and officers. In addition, the Delaware General Corporation Law, which forms the basis of the Courts' opinions, is widely regarded as giving great flexibility to corporations to manage their affairs. For these reasons, Delaware is considered to have the most business-friendly legal system in the United States; therefore a great number of companies are incorporated in Delaware, including 60% of the companies listed on the New York Stock

Exchange.[51] Delaware was the last US state to use judicial corporal punishment, in 1952.[52]

Executive branch

The executive branch is headed by the Governor of Delaware. The present governor is Jack A. Markell (Democrat), who took office January 20, 2009. The lieutenant governor is Matthew P. Denn. The governor presents a "State of the State" speech to a joint session of the Delaware legislature annually.[53]

Counties

Delaware is subdivided into three counties; from north to south they are New Castle, Kent County and Sussex. This is the fewest among all states, with Rhode Island having five counties. Each county elects its own legislative body (known in New Castle and Sussex counties as **County Council**, and in Kent County as **Levy Court**), which deal primarily in zoning and development issues. Most functions which are handled on a county-by-county basis in other states — such as court and law enforcement — have been centralized in Delaware, leading to a significant concentration of power in the Delaware state government. The counties were historically divided into hundreds, which were used as tax reporting and voting districts until the 1960s, but now serve no administrative role, their only current official legal use being in real-estate title descriptions.[54]

Politics

Presidential elections results

Year	Republican	Democratic
2008	37.37% *152,356*	**62.63%** *255,394*
2004	45.75% *171,660*	**53.35%** *200,152*
2000	41.90% *137,288*	**54.96%** *180,068*
1996	36.58% *99,062*	**51.82%** *140,955*
1992	35.33% *102,313*	**43.52%** *126,054*
1988	**55.88%** *139,639*	43.48% *108,647*
1984	**59.78%** *152,190*	39.93% *101,656*
1980	**47.21%** *111,252*	44.87% *105,754*
1976	46.57% *109,831*	**51.98%** *122,596*
1972	**59.60%** *140,357*	39.18% *92,283*
1968	**45.12%** *96,714*	41.61% *89,194*
1964	38.78% *78,078*	**60.95%** *122,704*
1960	49.00% *96,373*	**50.63%** *99,590*

The Democratic Party holds a plurality of registrations in Delaware. Until the 2000 presidential election, the state tended to be a Presidential bellwether, sending its three electoral votes to the winning candidate since 1952. Bucking that trend, however, in 2000 and again in 2004, Delaware voted for the Democratic candidate. In the 2000 election Delaware voted with the winner of the popular vote, Al Gore, who subsequently lost the Electoral College to George W. Bush. John Kerry won Delaware by eight percentage points with 53.5% of the vote in 2004. In 2008, Democrat Barack Obama defeated Republican John McCain in Delaware 62.63% to 37.37%. Obama's running mate was Joe Biden, who had represented Delaware in the United States Senate since 1973.

For a period of time the Republican Party was the dominant political party in the state due in part to the influence of the du Pont family. Ralph Nader assembled a working group to investigate ties between Delaware's politicians and industrialists and published its findings in the book *The Company State*.

The nominees of the Democratic Party have won the past four gubernatorial elections. Democrats presently hold eight of the nine statewide elected offices, while the Republicans hold only one statewide office, State Auditor.

Freedom of information

Each of the 50 states of the United States has passed some form of freedom of information legislation, which provides a mechanism for the general public to request information of the government. In 2011, Delaware passed legislation placing a 15 business day time limit on addressing freedom-of-information requests, to either produce information or an explanation of why such information would take longer than this time to produce.[55]

Government revenue

Delaware has six different income tax brackets, ranging from 2.2% to 5.95%. The state does not assess sales tax on consumers. The state does, however, impose a tax on the gross receipts of most businesses. Business and occupational license tax rates range from 0.096% to 1.92%, depending on the category of business activity.

Delaware does not assess a state-level tax on real or personal property. Real estate is subject to county property taxes, school district property taxes, vocational school district taxes, and, if located within an incorporated area, municipal property taxes.

Gambling provides significant revenue to the state. For instance, the casino at Delaware Park Racetrack provided more than $100 million USD to the state in 2010.[56]

Municipalities

Wilmington is the state's largest city and its economic hub. It is located within commuting distance of both Philadelphia and Baltimore. All regions of Delaware are enjoying phenomenal growth, with Dover and the beach resorts expanding at a rapid rate.

Further information: List of Delaware municipalities

Counties

- Kent
- New Castle
- Sussex

Cities

- Delaware City
- Dover
- Harrington
- Lewes
- Middletown
- Milford
- New Castle
- Newark
- Rehoboth Beach
- Seaford
- Wilmington

Towns

- Bellefonte
- Bethany Beach
- Bethel
- Blades
- Bowers
- Bridgeville
- Camden
- Cheswold
- Clayton
- Dagsboro
- Delmar
- Dewey Beach
- Ellendale
- Elsmere

Towns *(cont.)*

- Farmington
- Felton
- Fenwick Island
- Frankford
- Frederica
- Georgetown
- Greenwood
- Hartly
- Henlopen Acres
- Houston
- Kenton
- Laurel
- Leipsic
- Little Creek
- Magnolia
- Millsboro
- Millville
- Milton
- Newport
- Ocean View
- Odessa
- Selbyville
- Slaughter Beach
- Smyrna
- South Bethany
- Townsend
- Viola
- Woodside
- Wyoming

Villages

- Arden
- Ardencroft
- Ardentown

Unincorporated places

- Bear
- Brookside
- Christiana
- Clarksville
- Claymont
- Dover Base Housing
- Edgemoor
- Glasgow
- Greenville
- Gumboro
- Harbeson
- Highland Acres
- Hockessin
- Kent Acres
- Lincoln City
- Long Neck
- Marshallton
- North Star
- Omar
- Pike Creek
- Rising Sun-Lebanon
- Riverview
- Rodney Village
- Roxana
- Saint Georges
- Stanton
- Wilmington Manor
- Woodside East
- Yorklyn

Top 10 richest places in Delaware

Ranked by per capita income

1. Greenville: $83,223
2. Henlopen Acres: $82,091
3. South Bethany: $53,624
4. Dewey Beach: $51,958
5. Fenwick Island: $44,415
6. Bethany Beach: $41,306
7. Hockessin: $40,516
8. North Star: $39,677

Dover

9. Rehoboth Beach: $38,494

10. Ardentown: $35,577

Further information: Delaware locations by per capita income

Education

Delaware was the origin of *Belton v. Gebhart*, one of the four cases which was combined into *Brown v. Board of Education*, the Supreme Court of the United States decision that led to the end of segregated public schools. Significantly, *Belton* was the only case in which the state court found **for** the plaintiffs, thereby ruling that segregation was unconstitutional.

Unlike many states, Delaware's educational system is centralized in a state Superintendent of Education, with local school boards retaining control over taxation and some curriculum decisions.

As of 2011, the Delaware Department of Education had authorized the founding of 25 charter schools in the state, among them one all-girls facility.[57]

Newark

Seaford

Discipline

A study in 2010 indicated that while black students made up 32% of public school attendees, they accounted for 55% of the students who were suspended or expelled.[58]

Wilmington

Colleges and universities

- Delaware College of Art and Design
- Delaware State University
- Delaware Technical & Community College
- Drexel University at Wilmington
- Goldey-Beacom College
- University of Delaware
- Wesley College
- Widener University School of Law
- Wilmington University

Sister cities and states

Delaware's sister state in Japan is Miyagi Prefecture.[59]

Media

Television

There are no network broadcast-television stations operating solely in Delaware. A local PBS station from Philadelphia (but licensed to Wilmington), WHYY-TV, maintains a studio and broadcasting facility in Wilmington and Dover, while ION Television affiliate WPPX is licensed to Wilmington but maintains their offices in

Philadelphia and their digital transmitter outside of that city and an analog tower in New Jersey. Philadelphia's ABC affiliate, WPVI-TV, maintains a news bureau in downtown Wilmington. The northern part of the state is served by network stations in Philadelphia and the southern part by network stations in Baltimore and Salisbury, Maryland. Salisbury's CBS affiliate, WBOC-TV, maintains bureaus in Dover and Milton.

Tourism

While Delaware has no places designated as national parks,[60] national seashores, national battlefields, national memorials, or national monuments, it does have several museums, wildlife refuges, parks, houses, lighthouses, and other historic places. Delaware is home to the second longest twin span suspension bridge in the world, the Delaware Memorial Bridge.

Rehoboth Beach, together with the towns of Lewes, Dewey Beach, Bethany Beach, South Bethany, and Fenwick Island, comprise Delaware's beach resorts. Rehoboth Beach often bills itself as "The Nation's Summer Capital" because it is a frequent summer vacation destination for Washington, D.C., residents as well as visitors from Maryland, Virginia, and in lesser numbers,

Rehoboth Beach is a popular vacation spot during the summer months

Pennsylvania. Vacationers are drawn for many reasons, including the town's charm, artistic appeal, nightlife, and tax free shopping.

Delaware is home to several festivals, fairs, and events. Some of the more notable festivals are the Riverfest held in Seaford, the World Championship Punkin Chunkin held at various locations throughout the county since 1986, the Rehoboth Beach Chocolate Festival, the Bethany Beach Jazz Funeral to mark the end of summer, the Apple Scrapple Festival held in Bridgeville, the Rehoboth Beach Jazz Festival, the Sea Witch Halloween Festival and Parade in Rehoboth Beach, the Rehoboth Beach Independent Film Festival, the Nanticoke Indian Pow Wow in Oak Orchard, and the Return Day Parade held after every election in Georgetown.

Culture and entertainment

Sports

Club	Sport	League
Wilmington Blue Rocks	Baseball	Minor League Baseball
Delaware Griffins	Football	Women's Professional Football League
Central Delaware SA Future	Soccer	Women's Premier Soccer League
Delaware Dynasty	Soccer	USL Premier Development League
Diamond State Roller Girls	Roller Derby	Women's Flat Track Derby Association

As Delaware has no franchises in the major America professional sports leaguess, many Delawareans follow either Philadelphia or Baltimore teams, depending on their location within the state. The University of Delaware's football team has a large following throughout the state with the Delaware State University and Wesley College teams also enjoying a smaller degree of support.

Delaware is home to Dover International Speedway and Dover Downs. DIS, also known as the *Monster Mile*, hosts two NASCAR races each year. Dover Downs is a popular harness racing facility. It is the only co-located horse and car-racing facility in the nation, with the Dover Downs track is located inside the DIS track.

Delaware has been home to professional wrestling outfit Combat Zone Wrestling (CZW). CZW has been affiliated with the annual Tournament of Death and ECWA with its annual Super 8 Tournament.

Delaware is home to the Diamond State Games, an amateur Olympic-style sports festival. The event is open to athletes of all ages and is also open to residents beyond the borders of Delaware. The Diamond State Games were created in 2001 and participation levels average roughly 2500 per year in 12 contested sports.

Substance criminalization and abuse

According to a 2011 news publication, "abuse of prescription narcotics and anti-anxiety drugs has killed someone, on average, every other day over the last two years", with a death toll greater than that from either traffic accidents or abuse of traditional addictive substances (e.g. cocaine, heroin, alcohol).[61] Some of the prescription material reaches the market through theft from retail outlets, most are provided through a doctor's prescription.[61]

As of October 2011, Delaware criminalized "bath salts", placing it on the "state's list of banned substances".[62]

Delaware Native Americans

Delaware is also the name of a Native American group (called in their own language Lenni Lenape) that was influential in the colonial period of the United States and is today headquartered in Cheswold, Kent County, Delaware. A band of the Nanticoke tribe of American Indians today resides in Sussex County and is headquartered in Millsboro, Sussex County, Delaware.

Namesakes

* Several ships have been named USS *Delaware* in honor of this state.

See also

* List of places in Delaware
* List of Governors of Delaware
* National Register of Historic Places listings in Delaware

References

[1] "RESIDENT POPULATION OF THE 50 STATES, THE DISTRICT OF COLUMBIA, AND PUERTO RICO: 2010 CENSUS" (http:// 2010.census.gov/news/pdf/apport2010_table2.pdf). United States Census Bureau. . Retrieved 2010-12-21.

[2] "Elevations and Distances in the United States" (http://egsc.usgs.gov/isb/pubs/booklets/elvadist/elvadist.html). United States Geological Survey. 2001. . Retrieved October 21, 2011.

[3] Elevation adjusted to North American Vertical Datum of 1988.

[4] Schenck, William S.. *Highest Point in Delaware* (http://www.dgs.udel.edu/publications/pubs/factsheets/highestpoint.aspx). Delaware Geological Survey. . Retrieved 2008-07-23.

[5] *Random House Dictionary*

[6] While the U.S. Census Bureau designates Delaware as one of the South Atlantic States, it is sometimes grouped with the Mid-Atlantic States or the Northeastern United States.

[7] "Delaware" (http://www.etymonline.com/index.php?term=Delaware). Online Etymology Dictionary. . Retrieved 2007-02-24.

[8] (http://www.census.gov/Press-Release/www/releases/archives/voting/011400.html)

[9] "The State New Economy Index" (http://www.neweconomyindex.org/states/2002/delaware.html). The Progressive Policy Institute (PPI). 2002. .

[10] "Delaware Living History" (http://www.delawareliving.com/history.html). .

[11] "State of Delaware (A brief history)" (http://portal.delaware.gov/facts/history/delhist.shtml). .

[12] Because of surveying errors, the actual line is actually several compound arcs with centers located at different points in town New Castle

[13] Olson, D. M, E. Dinerstein et al. (2001). "Terrestrial Ecoregions of the World: A New Map of Life on Earth" (http://gis.wwfus.org/ wildfinder/). *BioScience* **51** (11): 933–938. doi:10.1641/0006-3568(2001)051[0933:TEOTWA]2.0.CO;2. ISSN 0006-3568. .

[14] Montgomery, Jeff (14 May 2011). "Cleaning up contamination" (http://www.webcitation.org/5ygBz8xTN). *The News Journal* (New Castle, Delaware: Gannett): DelawareOnline. Archived from the original (http://www.delawareonline.com/article/20110514/NEWS02/

105140360/-1/NLETTER01/Cleaning-up-contamination) on 14 May 2011. . Retrieved 14 May 2011. *The first online page is archived; the page containing information related here is not in the archived version.*

[15] John A. Munroe (2006). "3". *History of Delaware* (5, illustrated ed.). University of Delaware Press. p. 45. ISBN 0874139473. " Chapter 3. The Lower Counties On The Delaware (http://books.google.com/books?id=vs7NcOKnlNUC&lpg=PA45&dq="Lower counties" "on the delaware"&pg=PA46#v=onepage&q="Lower counties" "on the delaware"&f=false)"

[16] Lurie MN, Mappen M. Encyclopedia of New Jersey. Rutgers University Press, 2004, ISBN 0813533252, p. 327

[17] Mayo LS. John Wentworth, Governor of New Hampshire: 1767-1775. Harvard University Press, 1921, p. 5

[18] Simon Schama, *Rough Crossings: Britain, The Slaves, and the American Revolution*, New York, HarperCollins, 2006

[19] Paul Heinegg. Free African Americans in Virginia, North Carolina, South Carolina, Maryland and Delaware (http://www. freeafricanamericans.com/) Accessed 15 February 2008

[20] Peter Kolchin, *American Slavery: 1619–1877*, New York: Hill and Wang, 1994, pp.78, 81-82

[21] Peter Kolchin, *American Slavery: 1619–1877*, New York: Hill and Wang, 1994, pp.81-82

[22] Historical Census Browser, 1860 Federal Census, University of Virginia Library (http://fisher.lib.virginia.edu/collections/stats/ histcensus/), accessed 15 March 2008

[23] Dalleo, Peter T. (1997-06-27). "The Growth of Delaware's Antebellum Free African Community" (http://www.udel.edu/BlackHistory/ antebellum.html). University of Delaware. .

[24] Resident Population Data. "Resident Population Data - 2010 Census" (http://2010.census.gov/2010census/data/apportionment-pop-text. php). 2010.census.gov. . Retrieved 2011-08-17.

[25] "American FactFinder" (http://factfinder2.census.gov/faces/tableservices/jsf/pages/productview.xhtml?pid=DEC_10_PL_QTPL& prodType=table). Factfinder2.census.gov. 2010-10-05. . Retrieved 2011-08-17.

[26] "Population and Population Centers by State: 2000" (http://www.census.gov/geo/www/cenpop/statecenters.txt) (TXT). United States Census Bureau. 2002-02-20. . Retrieved 2007-03-09.

[27] SB 129 (http://legis.delaware.gov/LIS/LIS144.NSF/vwLegislation/SB+129?Opendocument), assigned 2007-06-13 to Senate Education Committee.

[28] HB 436 (http://legis.delaware.gov/LIS/LIS143.NSF/vwLegislation/HB+436?Opendocument), stricken 2006-06-15

[29] http://www.gc.cuny.edu/faculty/research_briefs/aris/key_findings.htm

[30] "State Membership Reports" (http://www.thearda.com/mapsReports/reports/state/10_2000.asp). *thearda.com.* . Retrieved 2010-06-15.

[31] "GDP by State" (http://greyhill.com/gdp-by-state). Greyhill Advisors. . Retrieved 9 September 2011.

[32] Eric Ruth (15 April 2010). "Delaware housing: Home prices slide in all three counties; sales in NCCo, Kent down from year ago" (http:// www.delawareonline.com/article/20110416/BUSINESS/104160310/-1/NLETTER01/ Home-prices-slide-in-all-three-counties--sales-in-NCCo--Kent-down-from-year-ago). *News Journal* (Gannett): Delaware Online. . Retrieved 16 April 2011.

[33] Luladey B. Tadesse (2006-08-26). "Del. workers earn 7th-highest salary in U.S." (http://www.liveinde.com/delawareno7salaries.htm). Delaware News-Journal. Archived from the original (http://www.delawareonline.com/apps/pbcs.dll/article?AID=/20060826/NEWS/ 608260340/1006/NEWS) on 2006-08-30. . Retrieved 2006-08-26. *Note: value of $937 per week was for the 4th quarter of 2005.*

[34] "Delaware Statewide Seasonally Adjusted" (http://data.bls.gov/pdq/SurveyOutputServlet?data_tool=latest_numbers& series_id=LASST10000003) (database report). *Local Area Unemployment Statistics.* United States Department of Labor, Bureau of Labor Statistics. February 2011. . Retrieved 16 April 2011.

[35] Jonathan Starkey (21 April 2011). "DuPont quarterly profit up 27%" (http://www.delawareonline.com/article/20110422/BUSINESS/ 104220333/-1/NLETTER01/DuPont-quarterly-profit-up-27-?source=nletter-news). *News Journal* (Wilmington, Delaware: Gannett): Business. . Retrieved 22 April 2011. "The company employs 8,000 people in Delaware and 60,000 globally."

[36] Eder, Andrew (2008-07-21). "DuPont can't avoid talk of buyout" (http://web.archive.org/web/20080801191131/http://www. delawareonline.com/apps/pbcs.dll/article?AID=/20080721/BUSINESS/807210335). *Delaware News-Journal* (Gannett). Archived from the original (http://www.delawareonline.com/apps/pbcs.dll/article?AID=/20080721/BUSINESS/807210335) on August 1, 2008. . Retrieved 2008-07-23. "Any type of takeover of DuPont -- the state's second-largest private employer, with about 8,900 employees -- would almost certainly mean local job cuts, said John Stapleford, a senior economist...."

[37] Jonathan Starkey (23 April 2011). "AstraZeneca to demolish three buildings" (http://www.delawareonline.com/article/20110423/ BUSINESS/104230343/-1/NLETTER01/AZ-to-demolish-3-buildings). *News Journal* (Wilmington, Delaware: Gannett): DelawareOnline. . Retrieved 23 April 2011.

[38] Barrish, Cris (10 July 2011). "Oversight of doctors improves" (http://www.webcitation.org/604podvh7). *The News Journal* (Newcastle, Delaware). ISSN 1042-4121. Archived from the original (http://www.delawareonline.com/article/20110710/NEWS02/110710003/ Oversight-doctors-improves) on 10 July 2011. . Retrieved 10 July 2011 *Note, only the first online page of the article has been archived.*

[39] "Delaware Division of Corporations" (http://www.corp.delaware.gov/aboutagency.shtml). Corp.delaware.gov. . Retrieved 2011-08-17.

[40] "Delaware 2007 Fiscal Notebook - State General Fund Revenues by Category (F.Y. 2002 - F.Y. 2005)" (http://finance.delaware.gov/ publications/fiscal_notebook_07/Section02/sec2page24.pdf) (PDF). . Retrieved 2011-08-17.

[41] Tax Justice Network (2009-11-01). "Financial Secrecy Index" (http://www.financialsecrecyindex.com/2009results.html). *financial secrecy index.* .

[42] Felicity Lawrence (2009-04-07). "Blacklisted tax havens agree to implement OECD disclosure rules" (http://www.guardian.co.uk/ business/2009/apr/07/g20-banking). *guardian.co.uk.* .

[43] "CHAPTER 7. REGULATORY PROVISIONS" (http://web.archive.org/web/20101106175332/http://delcode.delaware.gov/title4/ c007/index.shtml). *Online Delaware Code*. Delaware General Assembly. Archived from the original (http://delcode.delaware.gov/title4/ c007/index.shtml) on 6 November 2010. . Retrieved 13 September 2011.

[44] Aaron, Nathans (9 July 2011). "Del. package stores hope to benefit from Md. tax" (http://www.webcitation.org/605Q8t2S3). *The News Journal* (New Castle, Delaware). Archived from the original (http://www.delawareonline.com/apps/pbcs.dll/ article?AID=2011107100328) on 10 July 2011. . Retrieved 10 July 2011.

[45] Harlow, Summer (2008-01-20). "Auto tag No. 6 likely to sell for $1 million" (http://www.delawareonline.com/apps/pbcs.dll/ article?AID=/20080120/NEWS/801200351&template=printart). *The News Journal*. .

[46] "State of Delaware Department of Transportation" (http://www.deldot.gov/index.shtml). State of Delaware. . Retrieved 30 June 2006.

[47] Staff (Delaware Department of Transportation Public Relations) (2005). *Delaware Transportation Facts 2005* (http://www.deldot.gov/ information/pubs_forms/fact_book/pdf/2005/2005_deldot_fact_book.pdf). DelDOT Division of Planning. .

[48] "Projects: Delaware Bicycle Facility Master Plan" (http://www.deldot.gov/information/projects/bike_and_ped/bike_facilities/pages/ regional_routes.shtml). Delaware Department of Transportation. . Retrieved 2010-09-28.

[49] Justin Williams (17 April 2011). "Anything Once: On the road, taking plenty of pot shots" (http://www.delawareonline.com/article/ 20110417/NEWS02/304170008/-1/NLETTER01/On-the-road--taking-plenty-of-pot-shots). *News Journal* (Wilmington, Delaware: Gannett): DelawareOnline. . Retrieved 17 April 2011.

[50] Matzer Rose, Marla (9 January 2008). "Skybus adds two cities to schedule" (http://www.dispatch.com/live/content/business/stories/ 2008/01/09/skybus_routeannounce.ART_ART_01-09-08_C10_DP90M2P.html?sid=101). . Retrieved 2008-01-09.

[51] "Division of Corporations - About Agency" (http://web.archive.org/web/20070228002805/http://www.state.de.us/corp/ aboutagency.shtml). Delaware Division of Corporations. Archived from the original (http://www.state.de.us/corp/aboutagency.shtml) on 2007-02-28. . Retrieved 2007-03-09. *Note: replacement current URL (2008-07-23) is* http://www.corp.delaware.gov/aboutagency.shtml .

[52] Pleck, Elizabeth Hefkin (2004). *Domestic tyranny: the making of American social policy against family* (http://books.google.com/ books?id=zN2A2shTz6YC&pg=PA120). University of Illinois Press. p. 120. ISBN 9780252071751. .

[53] Staff (2010). "Home" (http://www.webcitation.org/5n1UxlPVx). *website for Delaware House of Representatives Minority Caucus*. Delaware House of Representatives Minority Caucus. Archived from the original (http://www.delawarestatehouse.com/) on 2001-01-24. . Retrieved 2001-01-24.

[54] "The Hundreds Of Delaware" (http://history.delaware.gov/museums/vc/vc_hundreds.shtml). *Department of State: Division of Historical and Cultural Affairs*. Delaware State Archives. . Retrieved 2010-09-28.

[55] Rep. Bennett; Sen. Peterson & Sen. Katz (6 January 2011). "10" (http://legis.delaware.gov/lis/lis146.nsf/vwlegislation/HB+5). *Online Delaware Code* (http://legis.delaware.gov/LIS/lis146.nsf/vwLegislation/HB+5/). **78** (published 15 April 2011). House Bill # 5. . Retrieved 22 April 2011

[56] Chris Barrish (23 April 2011). "Delaware crime: Wave of brazen attacks sounds alarm at casino" (http://www.webcitation.org/ 5ygCHfM0y). *News Journal* (Wilmington, Delaware: Gannett): DelawareOnline. Archived from the original (http://www.delawareonline. com/article/20110423/NEWS01/104230342/-1/NLETTER01/Wave-of-brazen-attacks-sounds-alarm-at-casino) on 14 May 2011. . Retrieved 23 April 2011. *First page of online article archived via link provided.*

[57] Dobo, Nichole (12 June 2011). "Delaware schools: Checkered past goes unchecked" (http://www.webcitation.org/5zOplxsjf). *The News Journal* (New Castle, Delaware). Archived from the original (http://www.delawareonline.com/article/20110612/NEWS03/106120369/ -1/NLETTER01/Checkered-past-goes-unchecked?source=nletter-news) on 13 June 2011. . Retrieved 13 June 2011.

[58] Dobo, Nichole (17 October 2010). "Black students punished more". Melbourne, Florida: Florida Today. pp. 13A.

[59] McDowell; Sen. McBride; Rep. George (22 March 2011). "Mourning Those Lost in the Recent Earthquake and Related Disasters that have Befallen Japan, and Expressing the Thoughts and Prayers of All Delawareans for the Citizens of Our Sister State of Miyagi Prefecture During These Difficult Times" (http://legis.delaware.gov/LIS/LIS146.NSF/vwlegislation/58CC58989B6361AD8525785B005381FB). Senate Joint Resolution # 3. . Retrieved 22 April 2011.

[60] "Delaware's historic role should be remembered" (http://www.delawareonline.com/article/20110430/OPINION11/104300305/-1/ NLETTER01/Delaware-s-historic-role-should-be-remembered). *The News Journal* (Wilmington, Delaware: Gannett): Opinion; delawareonline. 29 April 2011. . Retrieved 30 April 2011.

[61] Cris Barrish (7 November 2011), "Peddling Pain Pills" (http://www.delawareonline.com/article/20111107/NEWS1111/111070322/-1/ NLETTER01/Peddling-pain-pills), *The News-Journal*, , retrieved 8 November 2011

[62] Sean O'Sullivan (1 October 2011), "Drug criminalized in emergency ban" (http://pqasb.pqarchiver.com/delawareonline/access/ 2473070921.html?FMT=ABS&date=Oct+01,+2011), *The News Journal*, , retrieved 15 November 2011 *Only the abstract was available without purchase at the time of access.*

External links

- State of Delaware official website (http://www.delaware.gov/)
- Delaware travel guide from Wikitravel
- Delaware Tourism homepage (http://www.visitdelaware.com/)
- Delaware Map Data (http://datamil.delaware.gov/)
- Energy & Environmental Data for Delaware (http://tonto.eia.doe.gov/state/state_energy_profiles. cfm?sid=DE)
- USGS real-time, geographic, and other scientific resources of Delaware (http://www.usgs.gov/state/state. asp?State=DE)
- U.S. Census Bureau (http://quickfacts.census.gov/qfd/states/10000.html)
- Delaware State Facts (http://www.ers.usda.gov/StateFacts/DE.htm)
- 2000 Census of Population and Housing for Delaware (http://www2.census.gov/census_2000/datasets/ demographic_profile/Delaware/2kh10.pdf), U.S. Census Bureau
- Delaware (http://ballotpedia.org/wiki/index.php/Delaware) at Ballotpedia
- Delaware (http://judgepedia.org/index.php/Delaware) at Judgepedia
- Delaware (http://sunshinereview.org/index.php/Delaware) at Sunshine Review
- Delaware (http://www.dmoz.org/Regional/North_America/United_States/Delaware/) at the Open Directory Project
- Delaware State Databases (http://wikis.ala.org/godort/index.php/Delaware) - Annotated list of searchable databases produced by Delaware state agencies and compiled by the Government Documents Roundtable of the American Library Association.

gag:Delaware mrj:Делавэр

Article Sources and Contributors

Delaware_Route_44 *Source*: http://en.wikipedia.org/w/index.php?title=Delaware_Route_44 *Contributors*: Daniel73480, Dough4872, Imzadi1979

Kent_County,_Delaware *Source*: http://en.wikipedia.org/w/index.php?title=Kent_County%2C_Delaware *Contributors*: AXRL, Acntx, Ahoerstemeier, ArkansasTraveler, AshyLarry, Awiseman, Backspace, Boothy443, Bruno3469, Bumm13, CKirk92, CORNELIUSSEON, CanisRufus, Caponer, Chzz, Cindynunn, Citterio, CommonsDelinker, Corlier, D6, DCEdwards1966, Dadadata, Daniel Newman, Dough4872, Erasmussen, GoingBatty, Good Olfactory, Hailey C. Shannon, Hede2000, HennessyC, Hephaestos, Hyperfast, IsWayneBradygonnahavetosmacka, IsWayneBradygonnahavetosmackabitch, Isis, JW1805, Koyaanis Qatsi, Luxferro, Malepheasant, Mike mahaffie, Moncrief, Monegasque, Neier, Nyttend, Omnedon, Onyourside, Patricknoddy, PhilHibbs, Postdlf, PurpleChez, Qqqqqq, Ram-Man, Regulus, Revth, Rich Farmbrough, Robertjohnsonrj, RxS, Scott English, Sensor, Seth Ilys, Sfan00 IMG, Sgt Pinback, Smackhead12, Smallbones, Stilltim, Superman7515, Template namespace initialisation script, Timneu22, Tompw, WarthogDemon, WhisperToMe, WillC, Worldenc, Zoe, 15 anonymous edits

Delaware_Route_300 *Source*: http://en.wikipedia.org/w/index.php?title=Delaware_Route_300 *Contributors*: Dough4872, Imzadi1979, Jeff02, Matthiasb, Mitchazenia, NE2

Delaware_Route_8 *Source*: http://en.wikipedia.org/w/index.php?title=Delaware_Route_8 *Contributors*: Daniel73480, Dough4872, Imzadi1979, Rschen7754

Delaware_Route_11 *Source*: http://en.wikipedia.org/w/index.php?title=Delaware_Route_11 *Contributors*: Daniel73480, Dough4872, Imzadi1979, Iridescent

Hartly,_Delaware *Source*: http://en.wikipedia.org/w/index.php?title=Hartly%2C_Delaware *Contributors*: Acntx, Bumm13, Dough4872, Droll, Hmains, Nyttend, Pearle, Ram-Man, Rich Farmbrough, Robertjohnsonrj, Stilltim, VerruckteDan, 2 anonymous edits

Delaware_Department_of_Transportation *Source*: http://en.wikipedia.org/w/index.php?title=Delaware_Department_of_Transportation *Contributors*: Alansohn, Cheapestcostavoider, Dough4872, Dravecky, Edward, Exhartland, GoingBatty, Jshine20, KB3JUV, Mandarax, NellieBly, Neutrality, Overpush, Racepacket, Shearonink, Tangledorange, VerruckteDan, WhisperToMe, 5 anonymous edits

Delaware *Source*: http://en.wikipedia.org/w/index.php?title=Delaware *Contributors*: "Country" Bushrod Washington, 0, 100110100, 1973arierepjm, 1exec1, 7, A bit iffy, A8UDI, ABF, AC1, ARUNKUMAR P.R, Acather96, Achangeisasgoodasa, Achmelvic, Acroterion, Actingadam, Addison0426, AdjustShift, After Midnight, Agriculture8, Ahassan05, Ahoerstemeier, Aholcombe, Aivazovsky, Ajc trw, Aksi great, Aktsu, Alansohn, Alarics, Ale jrb, AlexCovarrubias, AlexiusHoratius, Alexwcovington, Alhutch, Alicatmeows, Aliceinlampyland, All Hallow's Wraith, Amazonien, Amb2132, Ampix0, Andiaswell, AndreNatas, Andrea105, Andy Marchbanks, Andy120290, Angr, Animum, Antandrus, Antonio Lopez, Antonrojo, Apartmento, ApprenticeFan, Aram33, Argo Navis, Argyriou, Arthena, Arthur3030, Asc85, AtherosKitten, AuburnPilot, Aude, AustinZ, Az1568, Aztom2, BD2412, BILZ, BSveen, Baa, Bachrach44, Bakedbeans1, Banaticus, BarretBonden, Bashen, Battleforkingdom, Bdodo1992, Beirne, Beland, Belinrahs, Ben Ben, BenBaker, Bender235, Benhealy, Benjh40, Bento00, Berean Hunter, Bhadani, Biasremover302, Big Mike 302, BillFlis, Billnye101, Binary TSO, Bkonrad, BlackRaspberry, Blacksabbath4343, Blanchardb, BoLingua, Bobblewik, Bobo192, Bogey97, Bolonium, Bomac, Bones789, BoondockArchitect, Boothy443, Braxiatel, BrendelSignature, Brendenhull, Brianga, Briangotts, Brion VIBBER, Brossow, BrownHairedGirl, Buaidh, Bwave, Bwgibson, CJLL Wright, CKirk92, Caesar Rodney, Caliga10, CanisRufus, Caoimhindebarra, Caponer, Carlaude, Carnifex, Casmith 789, Catgut, Cburnett, Cchow2, Ceyockey, Chargarther, CharlotteWebb, Chartran, Chasenm, Chaser, Cheapestcostavoider, Ched Davis, CheepnisAroma, Cheesypanda, Chensiyuan, Chester Markel, Cheung1303, ChrisCork, Christopher Mahan, Christopher Parham, Chun-hian, Cinefusion, Cinik, Civil Engineer III, ClairSamoht, Closedmouth, Cmdrjameson, Cmichael, Cntras, CombatCraig, Conn, Kit, Connormah, Conradthethird, Conversion script, CoolKid1993, Coolcaesar, Corinne68, Cornellrockey, Coulraphobic123, Cranes, Cribbswh, Crystallina, Cuchullain, Cumulus Clouds, Curlytomato, Curps, Cutelol23, Cybercobra, CylonCAG, Cypherrange, CzarB, DARTH SIDIOUS 2, DCEdwards1966, DLJessup, DXRAW, DabMachine, Dabomb87, Dae9577, DalKey, Dale Arnett, DancingPenguin, Danlaycock, Danny, Darkwind, Darth Panda, Dasajed, Daven200520, David Levy, DavidWBrooks, Dcornwall, Deflective, Dejayferri, Dekisugi, Delamed302, Delaware.gov, Delawarein, Delawareisawesome, Delirium, Denisarona, Deor, Der Sporkmeister, DerHexer, Descent, Devin Pederson, Diannaa, DigitalCatalyst, Dipics, Discospinster, DivineIntervention, Djlayton4, Dkdantastic, Docu, Dolovis, Dough4872, Doulos Christos, Dp462090, Dpodoll68, Dr. Blofeld, Dreadstar, Drmies, Drumlineramos, Dsmdgold, Dthomsen8, DuKot, Dukeofomnium, Dunganb, Dwayne, Dysprosia, ESkog, EWS23, EaglesFanInTampa, Eastlaw, Ebyabe, Eco84, Ed g2s, EdC, Eleigh33, Elpiseos, Emeraldcityserendipity, Emily120205, EncycloPetey, Engine13, Engineer Bob, English Bobby, Epbr123, Eopolk, Era One, Erfan1981, Ericenck, Erikmoore24, Ermeyers, Esrever, Etincelles, Eu.stefan, Evercat, Evice, Exhartland, Extransit, Extreme outdoors, Farosdaughter, FatTrebla, Faustus001, Fg2, Fieldday-sunday, Fifo, Finn-Zoltan, Flatterworld, Flewis, Flonto, FlugKerl, Flyers55, Flyguy33, Fobedra, Fogherty V. Tatin, Folajimi, Footballfan190, FrancoGG, Francs2000, Frazzydee, Frietjes, Fuhghettaboutit, Funandtrvl, Funnyhat, Future2008, GDonato, GHe, Gadfium, Gail, Gaius Cornelius, Galorr, GatosDeSilva, GcSwRhIc, Geologyguy, Germonger, GertVogel, Gilliam, Ginkgo100, Glane23, Gleppe, Gogo Dodo, Golbez, Goldfishbutt, Golgofrinchian, Good Olfactory, Gop 24fan, Gotthabieberfever, GraemeMcRae, Graham87, GrahamColm, Grazon, GregU, Griesfootball, Griffinofwales, Grunt, Gscshoyru, Gtstricky, Guaca, Guan, Guoguo12, Hadal, Haipa Doragon, HamburgerRadio, HappyCamper, Harris7, HarryHenryGebel, Hasek is the best, Hauganm, Hayfever, Haymaker, Hdt83, Heegoop, Helix84, HexaChord, Highvale, Hike395, History21, Hitchhiker89, Hmains, HokieRNB, Hoppingalong, Horologium, Hottentot, Howardjp, HowiePerlman, Hroðulf, Htconner, Htgrgwwew, Hungrybilly, Hunterdworsky, Hushpuckena, Hut 8.5, HyLander42, I already forgot, IBook of the Revolution, Iain99, Ibagli, Icehcky8, Igo4U, Igoldste, IhateMileyCyrusforever, Imperial Star Destroyer, Invincibleone, Isis, Ismbuff, Isomorphic, J.delanoy, JForget, JFreeman, JGCollins, JHMM13, JW1805, Jahanas, Jakabcin, Jake Wartenberg, JamesBWatson, Jamesontai, Jaxl, Jay, Jcam, Jengod, Jeppler, Jeremiestrother, JethroElfman, Jevansen, Jfpierce, Jhendin, Jiimmbob, JimIrwin, JimVC3, Jimmy678910, JinJian, Jj137, Jmsam81, Jmundo, JoSePh, JoeSmack, John K, JohnCub, Johnandsandie, Johnnywl, Jollyanswer, Jonathunder, Jose77, Josh3580, Journalism12, Jpg, Jptdrake, Juantay, Judeeclare, Juliancolton, Jultemplet, Justin1920, Justinfoster6961, Juzeris, K3vin, Kabri, Kahuroa, Kaishmoney2, Karmafist, Katalaveno, Kathie48, Katieh5584, Kayleigh66, Keegscee, Keilana, Keith Edkins, KeithH, Kelestar, Ken g6, Kenguest, Khan3817, Kimchi.sg, Kintetsubuffalo, KnowledgeOfSelf, Kozuch, Kramden4700, Kristinpedia, Kukini, Kurykh, Kwamikagami, Lamb Kebab, Laurascudder, LaytonGuy, Laytonmg, LcawteHuggle, LeaveSleaves, Ledcraft, LeeG, Legoktm, Leuko, Levelhead, Levineps, Leviton, Liechten11, Lightmouse, Lil chris brown, Lilac Soul, Lindaadnil, Lithpiperpilot, LittleChu542, Littledrummrboy, Littlesmalls, Logan, Loodog, Looxix, Lord Voldemort, Lozeldafan, Lst27, Ludalex, Luna Santin, M C Y 1008, MER-C, MJ94, MJCdetroit, MPerel, Magister Mathematicae, Maher-shalal-hashbaz, Manderiko, Manuel Trujillo Berges, Maproom, Marcusscotus1, Mareino, Marek69, Markaci, Mars11, Mastercampbell, MattieTK, Mav, Maxim, Maxis ftw, Mbvigil, Mckibkai000, MeddmaWamm, Mephistophelian, Mgard7331, Mglan, Michael Devore, Michael Hardy, Michael Schubart, Michaelschmatz, Miguel.v, Mike mahaffie, Mike p 8, Mike s, MikeLynch, Mild Bill Hiccup, Miquonranger03, Mirv, MisfitToys, Miskwito, Mister Pinball, Misterrick, Mkativerata, Mliggett, Mmahaffie, Modster, Moonriddengirl, Moorematthews, Morwen, Moverton, Mr. Lefty, MrIntellect, MrPMonday, MrRadioGuy, Mrdavidjb, Mredurk, Mrschimpf, Mu Cow, Mxn, MySuperiorInEveryWay, Mygerardromance, Mysid, N5iln, NE2, NEICenergy, NICKRICK, Nakon, Ncpanthers24, NellieBly, Neutrality, NewEnglandYankee, Newman6000, Nick123, Ninetyone, Ninly, Nkrosse, Nlu, NorwegianBlue, NovaDog, Nphs2010, Numbo3, Nwbeeson, Oda Mari, Ohnoitsjamie, Okedem, Old Guard, Olude, Onorem, OrweL2, Oscarthecat, Oxymoron83, PAWiki, PDH, Pannekaker, Parkwells, Patent.drafter, Patrickesmonde, Patricknoddy, Pavel Vozenilek, Pekinpekin, PeruvianLlama, PeterSymonds, Phatboi858, Phil Boswell, Philip Trueman, PicoGils, Pinball22, Pinethicket, Pinkadelica, Pizzacheeserules, Pjbflynn, Plasticup, Pogo-Pogo-Pogo, Polker03, Portillo, Postdlf, Pras, Prashanthns, Prattflora, PseudoSudo, Psycho Kirby, Ptolemy Caesarion, Pugsrock9, Pugwawa, Pwt898, Pxma, Qqqqqq, Quintote, Qworty, Qxz, R'n'B, R.T.Gellar, RL0919, RUL3R, RandomP, RandomRandomz, Raul654, Ravedave, Rboatright, Rdiedam, Realm of Shadows, Reaper Eternal, Reconfirmer, Reedy, Reesh, Regibox, Res2216firestar, Retired username, Retro00064, Revas, Rfc1394, Rich Farmbrough, Richard Getz, Rick Block, Rickard Vogelberg, Rivertorch, Rjensen, Rjt1000, Rjwilmsi, RoadieRich, Rob Hooft, Romanm, Root Beers, Rossami, Roxie524, Roxya, Rreagan007, Rreuling, Rtarpine, Ruhrfisch, RyanCross, Rzf3, SCEhardt, SD6-Agent, SMC, SNIyer12, Salamurai, Sam Korn, Sammy snake, Samwaltz, Sanfranman59, Sango123, Saopaulo1, SatyrTN, Saumoarush, Savant13, ScaldingHotSoup, Scanlan, Scapler, Scarian, Schzmo, Scientizzle, ScottyBerg, Scoutersig, Seattlenow, SebRovera, Sebh007, Sedna10387, Sensor, Sfbayroad, Sfitzpatrick924, Sfmontyo, Shadow Android, Shadowjams, Shalom Yechiel, SheepNotGoats, ShelfSkewed, Shelleyaug6, Shelofdel, Shenme, Silivrenion, Sir rupert orangepeel, Sjakkalle, Skeithkiller, SkerHawx, Skier Dude, Skizzik, Skookum1, Slakr, Slysplace, Smackhead12, Smalljim, Smallman12q, Sman24, Smashmint, Smeshy, Smnc, SmthManly, Sn00kie, Snowdog, Solipsist, Soonstory, South Bay, Southern Illinois SKYWARN, SpK, Spandox, SpartanPhalanx8588, Spellcast, Spettro9, Spikebrennan, Splenditello, Spoojmasta, Spyder130, Spydrlink, Srich32977, StAnselm, StanZegel, Stay cool, Stealthbreed, Stephen G. Brown, Stepp-Wulf, SteveSims, Steven Weston, Stillhere, Stilltim, Student7, Subverted, Superbeecat, Superman7515, Surmur, SuzukaISichigo100%, Swollib, Swpb, Symphony Girl, THEN WHO WAS PHONE?, TUF-KAT, Tachin34, Tarkan1st, Tbhotch, Tdadamemd, Tedickey, Telso, Template namespace initialisation script, Tgeairn, Thatguyflint, The Epopt, The Thing That Should Not Be, The Transhumanist, The-bus, Thekrashkid, Themfromspace, Theresa knott, Thingg, ThisisforCD, Thuresson, Thw1309, TicketMan, Tide rolls, Tidying Up, Tim1357, Timneu22, Tins128, Tinton5, Tombomp, TommyBoy, Tony1, Torahjerus14, Tpbradbury, Tracer9999, Train2104, TransUtopian, Trimalchio, Triona, Troy88, True Pagan Warrior, TruthIIPower, Turgan, Turover, TwinCityIL, Tyger63, Tyler, Uncle Dick, Unschool, Utcursh, Valerius Tygart, Vampire909, VanillaBear23, Vanished User 1004, Veinor, Velvetron, Verrai, Viriditas, Vpuliva, Vrenator, Vsmith, WOSlinker, Waggers, Wallsaregreat, Wangi, Wapcaplet, Wars, WarthogDemon, Washbummav, Wavelength, Weatherman78, Wendy23c, WhereAmI, Wheresamber, WhisperToMe, Wiki alf, Wiki-princess95, Wikieditor06, Wikipe-tan, Wikipelli, Wildwill2002, Will Beback, WilliamJE, Willking1979, Willydick, Wkharrisjr, Wknight94, Wmahan, Womble, Woohookitty, Wrath of Roth, Wrightbus, Wytukaze, XRK, Xcom, Xyzzyva, Yayme23, Yekrats, YellowMonkey, Yintan, Youngamerican, Zach4636, Zereshk, Zhou Yu, Zima3000, Zyxw, Zzedar, Zzuuzz, Zzyzx11, 1712 anonymous edits

Image Sources, Licenses and Contributors

Printed by Books on Demand GmbH, Norderstedt / Germany